UNDERSTANDING
IRAQ

ALSO BY WILLIAM R. POLK

Backdrop to Tragedy:
The Struggle for Palestine

The Opening of South Lebanon, 1788–1840

The United States and the Arab World

Passing Brave

The Elusive Peace: The Middle
East in the Twentieth Century

The Arab World Today

Neighbors and Strangers: The Fundamentals
of Foreign Affairs

Polk's Folly: An American Family History

UNDERSTANDING
IRAQ

*The Whole Sweep of Iraqi History,
from Genghis Khan's Mongols
to the Ottoman Turks to the British
Mandate to the American Occupation*

WILLIAM R. POLK

HarperCollins*Publishers*

HarperCollins books may be purchased for educational, business, or sales promotional use. For information, please write: Special Markets Department, HarperCollins Publishers Inc., 10 East 53rd Street, New York, NY 10022.

FIRST EDITION

Designed by Christine Weathersbee

Printed on acid-free paper

Library of Congress Cataloging-in-Publication Data is available upon request.

ISBN 0-06-076468-6

05 06 07 08 09 ❖/RRD   10 9 8 7 6 5 4 3 2 1

To the memory of Nadia Younes, dear friend,
killed on duty for the United Nations,
Baghdad August 19, 2003

AND

For Adib el-Jadir, my first guide to Iraqi politics,
a man of great principles who paid dearly
for them under Saddam

AND

For Hume Horan, longtime colleague,
friend, and former student, who sought
to find a way to bring peace to Iraq.

CONTENTS

MAPS

PREFACE

As an author, I have always felt that the reader deserves to know what the author is trying to do and how he proposes to do it.

Here I am trying to give as complete a "portrait" of Iraq as is now possible so that readers can evaluate the often confusing daily events. A greater understanding is necessary if we are to seek a safer, saner, and more humane future for the people of this troubled, wounded society and a less dangerous, less costly, and more productive relationship with them. Beyond Iraq itself, the impact of events there will shape much of what Americans think about and do for years to come in Africa and Asia. *Understanding Iraq* is the first step toward comprehending the not-so-brave new world that we and our children will face.

How to understand Iraq? As a historian, I believe that knowing about events-over-time is crucial to a perception of the present. I have been influenced by archaeologists to go back to the beginnings rather than merely intercepting events somewhere along their trajectory. And from anthropologists, I have learned to dig below actions to discover their social dimensions. So the reader will find that this book is cast widely and deeply to put contemporary issues and events in their fundamental or causative historical context.

I have been at this for more than half a century. I first went to Iraq in 1947, lived there (as a fellow of the Rockefeller Foundation) during 1951–52, and have returned dozens of times; I have visited almost every nook and cranny of the country; and, speaking their language, I have held long discussions with countless Iraqis. I studied Arabic and Turkish language and literature at Oxford and taught history, politics, and Arabic language and literature at Harvard and the University of Chicago. Finally, I was responsible for planning American policy toward the Middle East during the Kennedy administration.

What the reader will find is a distillation of half a century of reading and research, my own personal appreciation of the Iraqis over the long reach of history, and my best guesses on the future. Both as a historian and as a planner of government policy, I tremble at Hegel's admonition that "Peoples and governments never have learned anything from history or acted on principles deduced from it." Let us pray that we can prove him wrong. Otherwise, as the American philosopher George Santayana warned, we shall be doomed to repeat it.

ACKNOWLEDGMENTS

In more than half a century of living in, visiting, reading about, and discussing events in Iraq, four years of planning American policy on Iraq and many months of writing essays and books on Iraq, I have incurred debts to many friends, colleagues, teachers, and students in Iraq, America, Britain, Russia, and elsewhere. I cannot name them all but I must name a few: Costa Halkias, my first host in Baghdad; Adib al-Jadir, my first guide to Iraqi affairs; Sumaiyah Zahawi, who in 1952 illuminated a then-new dimension of Iraq for me; Michael Adams, the late Howard K. Smith, the late Murray Kempton, Neil Sheehan, John Cooley, Saïd Aburish, Jonathan Randal, Eric Rouleau, Peter Scholl-Latour, and Charles Glass were among the outstanding journalists it has been my privilege to count as friends. From them and, above all, from my brother, George Polk, I have received early and constant inspiration. To Sir Hamilton Gibb, Sir Edward Evans Pritchard, Emrys Peters, Albert Hourani, and Chaim Rabin I owe unmeasurable debts for teaching facts, techniques, and inspiration. Among fellow historians, I can name but a few from whose works on aspects of this book I have profited, Jacques Berque, Moshe Ma'oz, Charles Issawi, Philip Ireland, Steven Longrigg, Frank Stoakes, Majid Khadduri, Gabriel Baer, Hanna Batatu, Yitzhak Nakash, and Phoebe Marr. Among government officials, I have shared opinions, agreements, and arguments with Evgeni Primakov, McGeorge Bundy, Walt Rostow, Chester Bowles, Hume Horan, James Spain, Thomas Hughes, Lord Caradon, U Thant, Sir David Gore-Booth, Zaid ar-Rifai, and Nadia Younes. None of the above, of course, should be blamed for any mistakes I may have made.

My editor, Cass Canfield, Jr., and my literary agent, Sterling Lord, have been supportive throughout the task of writing. The generosity of William Polk Carey made the devotion of my time to the project possible. Last, I am greatly indebted to my wife, Elisabeth, and to my children, Milbry, Alison, George, and Eliza, and to my longtime dear friend Cooper Blankenship, for their continued support and affection through this literary venture and other trials and tribulations.

A NOTE ON WORDS
AND SPELLINGS

One of the tragic features of the American invasion and occupation of Iraq is the Iraqi and American lack of understanding of what each other means. At the most dramatic, this has resulted in many deaths; more pervasive has been the spread of suspicion and hatred among both peoples; but beyond these is a problem of even greater and more long-term importance: while we would scoff at a person who came to our country to report on us without being able to understand what we say, we must rely for what we learn about Iraq primarily on American officials and reporters who do not know the local language. Consequently, misunderstandings abound in their accounts. Arabic is a difficult language, requiring years of study; coming from a "family" unlike English, Semitic rather than Indo-European, it is organized in a dissimilar way and its concepts form associations that are not easy for a foreigner to grasp. Moreover, it is an ancient language, embodying about two thousand years of literature in a form intelligible to modern readers primarily because it was "frozen" in the Quran and in the poems that all school children memorize. So, beyond dictionary "meanings" of individual words loom literary allusions, associations and memories that are fundamental to modern thought.

In part these associations are "rooted" in basic meanings. As in Hebrew and other Semitic languages, the "root" of a word, usually three or four consonants, is elaborated by a set pattern of morphological changes. These changes enable a simple concept to take on quite complex meanings in all of which the basic concept remains. Since English and French, for example, do not have such internal relationships, translation often loses what is significant to the native speaker. This becomes clear when we consider how we would regard a foreign journalist coming to America to report on American politics who could not speak English and knew little or nothing about American society and culture.

With these thoughts on language in mind, consider four words that recur frequent in today's press, "martyr," "holy war," "religious leader," and "reactionary."

The word we translate as "martyr" is based on the root *Sh-H-D*. (*Sh* is one letter in Arabic.) The root idea is "to witness," *ShaHaDa*. From this comes the notion of giving evidence. To give evidence, a person swears, *ShaHaDa*. He may also affirm his belief in the Islamic faith, *taSHaHHada* by saying "There is no God but God . . . ." And from this comes the notion of suffering the consequences of having testified or affirmed one's faith, martyrdom, *ShaHāDa*. From this elaborate skein of ideas and emotions comes what is referred to in the press as a suicide bomber, *muShtaHiD*, one who seeks martyrdom. For the Muslim, this has little or none of the opprobrium conveyed by the English words. It is an act, indeed the supreme act, of one who is bearing witness to God of his faith.

The word we translate as "holy war," *JiHaD* derives from the root *JHD*. The basic meaning is to "work hard" or "strive." So *aJHaDa* is to overload (a camel) or exhaust (a person). Life can become hard, *JuHiDa*. One can drive another forward, *aJHaDa*, or one can drive himself, *taJaHaDa*, or seek to get to the very bottom of things, *istaJHaDa*. So mixed together are concepts of ability, power, energy, zeal, and fatigue. A *muJtaHiD* is a scholar who through long study seeks to get to the bottom of things, a jurisconsult of Islam, while a *muJaHiD* is a militant. That word was even used in Egypt as a military rank, a sergeant. A related noun, *maJHuD*, became the word for electric current. And, finally, of course, *JaHaDa* means to fulfill religious duty, including, if required, fighting for the faith against infidels.

What we term a "religious leader" is divided in Arabic into two very different concepts. As described above, a *muJtaHiD* is a scholar who is revered for his profound knowledge. He may be honored as a pious man, but it is his knowledge, acquired by years of study, that sets him apart. Thus, his standing is personal. Very different is another kind of religious leader, an *imam*. The basic meaning of *imam* is a person who stands in front. This is the task of a leader of prayer who sets an example for those who assemble with him in reciting and genuflecting. So important was this function in early Islam that the man who performed it became the caliph at the death of the Prophet Muhammad. Later the term was applied by the Shia Muslims to those descendants of the Prophet who were thought to have inherited in some mystical way the "spirit" of God. Thus, the martyr of Shiism, Husain, a grandson of Muhammad, is revered as the *imam*, the preeminent figure of his faith.

The fourth concept is in many ways the most complex and interesting; it grows out of the Arabic root *SLF*. Confusingly to foreigners, *SuLaF* can mean both to be past and to be future, to be the vanguard and the rear guard. How this is possible cuts to the heart of thinking about religious politics. Like the European and American Puritans, Muslim thinkers have sought to "purify" or return to the fundamental concepts of their forefathers, the *aSLaF*, ridding their religion of accretions that they feel have perverted it. So the *as-SaLaFiyah* movement, inspired by a great

nineteenth-century Egyptian theologian, has spread from
Morocco to Iraq. Its adherents do not think of themselves as
"reactionary" any more than European Puritans did; rather, they
see themselves as the vanguard, reformers, progressives. They
bring out the pure essence (the same root gives *SuLaF*, the finest
juice of fruit); they enable the produce of religious literature to
be reaped by a *miSLaFah* (a harrow) and they nourish (*SaLLaF*)
the belief of their companions.

The problems of understanding peoples who speak other
languages, hold other faiths, and are guided by other customs
make simple words both obscure and illuminating. Seeking to
know precisely and exactly what people mean in Arabic requires
years of study; absent that study, the observer skates on the surface without really perceiving what other people mean. Readers
of the press and observers of government action will see many
examples of that misinterpretation. Misunderstanding is not just
an academic or pedantic issue: it has already cost many lives in
Iraq. *Understanding Iraq* thus begins with words.

Arabic words do not transliterate easily into English.
Increasingly, as scholars on Chinese also have done, Arabists
have evolved standards. Some merely confuse a person who does
not know Arabic. So I keep very familiar words like "Mecca,"
rather than the more correct Makkah. Long ago, the style of tag-
ging the names of dynasties with the letters "ad" or "id" was set;
so Umayyad, Abbasid, etc. They are too familiar to tamper with.
But where the words are names or are not familiar, I use
spellings that more correctly catch the Arabic sound and writing: the Iraqi general I spell as Qasim rather than the newspaper
version Kassem, and the dictator Saddam Husain rather than
Hussein. The sect of Islam is the *Shia* [Arabic: party (of the
Imam Ali)] and its adherents or partisans are *Shiis.*

INTRODUCTION

The first question a reader might ask is "Why is Iraq important?" Like all simple questions, this one has complex answers. They depend partly on who is asking, when he is asking, and what he has in his mind. I begin with Americans right now.

Since February 2003, the American invasion and occupation of Iraq has cost America more than one thousand lives, five thousand seriously wounded, and roughly $200 billion, and has adversely affected American interests throughout the world. The costs are not over: the total monetary cost may rise to at least half a trillion dollars. More Americans will be killed or wounded. How many is anyone's guess. And most observers believe it will take many years to rebuild what America has lost in what President Eisenhower called "the decent respect of mankind." While respect is ephemeral, it is not trivial. Nations depend as much or more on what has been aptly termed their "soft power" as on their economic or military might. The respect, one is tempted to say the "love," others have felt toward America has been traditionally one of the country's most valuable assets. Finally, as the scandal of the torture of prisoners, the flouting of international law, and attacks on American civil liberties have demonstrated, this war, like all violent conflicts, has eroded America's most prized attribute; a national character that has arisen from belief in liberty, justice, and decency.

Americans by and large thought that a war against Iraq was justified. They were told by their government that Iraq had weapons of mass destruction, was planning an attack on the United States, was actively supporting the terrorists who had attacked the World Trade Center and the Pentagon. When these charges proved to be untrue, the Bush administration insisted that a more significant reason was the foul and tyrannical nature of the Iraqi regime. That charge was true. But it was obviously not unique to Iraq. Several regimes, including some staunchly supported by the American government under both Democratic and Republican administrations, have engaged in horrific abuses of their citizens. Nor, almost everyone now concedes, was this the reason for the American attack.

More probable reasons have been openly discussed both in America and abroad. Included among them is that Iraq is immensely rich in oil and Iraqi oil is the cheapest in the world to produce. Securing the flow of oil from the Middle East on acceptable terms has been a fundamental American government objective for half a century. When the decision-making process of the Bush administration is finally revealed, it is likely that oil will figure prominently. As Undersecretary of Defense Paul Wolfowitz proclaimed at the Asia Security Summit in Singapore on June 3, 2003, Iraq was known to be "swimming" in oil. As the costs of the war have risen, as the announced objectives of American policy have seemed unlikely to be attained, and as statements made by the government have proved false, public opinion polls indicate that many Americans have come to distrust their government. This too must be figured as one of the costs of the conflict.

• • •

For Iraqis, the conflict takes on quite a different cast. For them, the costs and the benefits are not purely statistical. Statistics, however, are impressive. In the first phase, the actual invasion and aerial strikes, at least ten thousand Iraqis died and perhaps twice that many were greviously injured; property damage will almost certainly exceeed $200 billion as a result, primarily, of a bombing campaign more intense than any country had ever suffered short of nuclear weapons. Less evident are various other effects of a decade of sanctions and two years of occupation. A generation of children has grown to maturity without the nutrition and medical attention their fathers and mothers had; perhaps half a million have been permanently harmed; patterns of life were aborted or transformed; that abstract but real sense of national "honor" has been affronted; the mechanisms of law and order, imperfect though they certainly were, have been overturned; and a whole generation has lost crucial years in its development.

What many would assert they have gained is an end to the tyranny of Saddam Husain, although some, having seen examples of barbarity in the treatment of prisoners, now believe that the change is not so complete as they, and we, have proclaimed. Others worry, as I also do, that the current period without a dictator may turn out to have been only an interval between this dictator and the next. Some even believe that Saddam, if let out of prison, could reemerge.

What the many conflicts since the Second World War should have taught us is that warfare is brutalizing. Habits learned during struggle and justified by it are hard to drop even when peace returns. As I shall document, Iraq has suffered from a long history of violence; even in periods of relative peace, it had little experience with constructive civil order, and today, it is engaged in a struggle against the foreign invaders and between those of its citizens who are willing to work with us and those who are not. In this sense also, what is happening in Iraq may be a template for its future or, conceivably, for other countries if the American crusade continues.

Rebuilding Iraq may provide it with better facilities to replace what the bombing destroyed. That many Iraqis do not value the facilities above their independence is shown by the ferocity of their attacks on both the new facilities and on those seeking to build them. No different from other peoples, Iraqis show an evident propensity to place a higher value on independence than on things, even if those things are essential for a better way of life.

That better way of life will ultimately depend on the emergence of some form of representative, mutually tolerant government based on a modicum of respect for the rule of law. Whether they will do so under American occupation or influence is at least questionable. One thing is sure: they will emerge only by internal developments; they cannot be imposed by foreigners. A lesson of the past is crucial here. The Iraqi experience in what I have called "British Iraq" and "Revolutionary Iraq" demonstrates that the very *concepts* of representative government and the rule of law themselves need protection. If they are debased by being ascribed to unrepresentative, illegal, or undemocratic practice, they will again be, as they were then, crippled at birth.

For the world as a whole, the Iraq war has ushered in a major transformation: as Egyptian president Hosni Mubarrak wisely warned, it has created a hundred bin Ladins while ostensibly being fought to destroy the original one. It has made the task of terrorists easier because it has created large numbers of disaffected people in areas where previously there was no terrorism. It has pulled resources away from the struggle against hunger, drug abuse, disease, and protection of the environment toward what is euphemistically called "security." And it has diminished respect for law, justice, and freedom. So events in and around Iraq have rippled across the world, from Latin America to Indonesia; from Central Asia to South Africa; from Spain to the Philippines.

The second question a reader might ask is *"What* is Iraq?"

Consider first Iraq as a state. Much has been made of the statement that Iraq is "artificial." That is true. Most countries either still are or were until recently. Few can be dated for more than a century in something like their present form. Few are homogeneous. Consider China (with fifty-six separate "nations"), Russia (even after the breakup of the Soviet Union with at least a score), India (with scores in addition to its component states), and Indonesia (with almost a thousand). All existing African states, having been created by European powers to fit their convenience, are "artificial." Special cases? No, even such stable, historic states as France and Spain are now being forced to recognize their multinational character.

Iraqis lived for centuries under the Ottoman Empire and for millennia under various other regimes. I will discuss these formative experiences. Briefly, Iraq became a state at the end of the First World War, not by its own actions but as guided by the British government. It was composed of three parts, each formerly a province of the Ottoman Empire, in which lived Indo-European (Kurdish)–speaking Sunni and Shia Muslim Kurds in the north, Semitic-speaking (Arabic) Sunni Muslim Arabs in the center, and a combination of Arabic-speaking Sunni Muslim Arabs and Arabic-speaking Shia Muslim Arabs in the south. These peoples have lived together as a nation-state from roughly 1921 to the present. Their relationships have not been harmonious or stable, but, for the most part, they found them more acceptable than available alternatives. They have learned that when outsiders or they themselves emphasize their differences, they often pay a terrible price for failing to find common causes.*

The land of Iraq, like the people, is diverse but unified. The unity arises from two features: the river system of the Tigris and Euphrates and the fact that without them almost all Iraq, except for the north and part of the east, would be desert because it receives fewer than 8 inches, or 20 centimeters, of rainfall (the amount required to sustain agriculture) in an average year.

The total area of the country, as constituted in 1921, is about 172,000 square miles, or 437,065 square kilometers. That is slightly larger than California and not quite two-thirds the size of Texas.

Along the northern and eastern fringes of the country, making up about 5 percent of the total area, are mountains. The Zagros mountains form a band along the Turkish and Iranian frontiers about 250 miles, or 400 kilometers, long and 125 miles, or 200 kilometers, wide. Relatively well watered, the Zagros was the seedbed for the agricultural revolution that made possible the growth of early Iraqi civilization.

---

* There are roughly 16 to 20 million Arabs in Iraq, of whom 65 to 80 percent are Shiis and 20 to 30 percent are Sunnis. Kurds number roughly 3.6 to 4.8 million and Turkomans and others roughly 2 million.

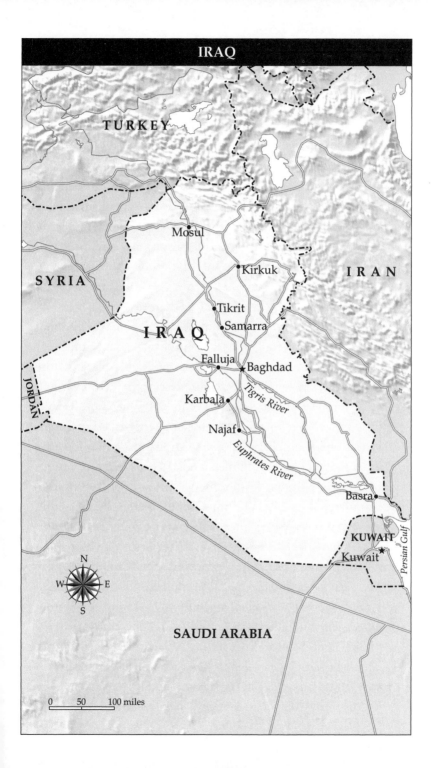

Higher (with one peak reaching nearly 12,000 feet, or 3,607 meters) and cooler than the plain, the rugged hillsides and deep valleys have supported and sheltered the vigorous, warlike, and independent peoples whom we know today as the Kurds. Not all Kurds live in Iraq; they make up about one in ten Iranians, one in nine Syrians and one in five inhabitants of Turkey. Considered as a country, Kurdistan, the land of the Kurds, is larger than Iraq, about 200,000 square miles, or 520,000 square kilometers. That is roughly the size of California and Pennsylvania combined. The extent of Kurdistan, its diversity, its failure to achieve statehood, and its location have been and remain powerful influences on Iraq and, indeed, on the whole Middle East.

In the center of the country is a flat plain that makes up about a quarter of Iraq's total surface, about 47,000 square miles, or 121,500 square kilometers. This area is naturally (that is, before some of it was irrigated) a subtropical desert with extremely hot, rainless summers. About a quarter of its soil is theoretically suitable for agriculture; however, because of intense solar heat, evaporation is rapid, and, because of poor drainage, salting is everywhere a problem and south of Baghdad has been, since history was recorded, a life-or-death struggle. The south, Iraq's sump, was until recently a vast swamp. At the bottom of the south, where the rivers come together, is the outlet to the Persian Gulf.

The life of Iraq depends on the river systems. Only about 12,500 square miles, or 32,375 square kilometers—roughly the size of Massachussetts and Connecticut combined—can be farmed with rainfall; every other bit of agricultural land must be fed from the rivers. The Euphrates rises in Turkey and flows through Syria; less than half of the water rises in Iraq. When it reaches Baghdad, it is about the size of the Arkansas River at Little Rock. The Tigris, most of whose water rises in the Iraqi mountains, is about as large as the Missouri River at Kansas City. These rivers and their tributaries have traditionally made it possible to grow about enough winter wheat and barley, plus some vegetables, rice, and dates, to feed the small population; more recently, Iraq has had to import most of what it consumed. Thus, despite the image of Iraq as the "Garden of Eden," it is a poor country.

Poor, that is, on the surface. Beneath the surface are a number of pools of oil that, in the aggregate, are probably the world's largest. The first field was developed at Kirkuk in the Kurdish north. Other fields were developed in the south after the Second World War. Still undeveloped is a vast sea of oil believed to be under what has recently been nicknamed "the Sunni triangle" around Baghdad. That one field may equal all of Saudi Arabia's fields. Oil has been both the curse and the blessing of Iraq: curse because it has excited the greed of others, and blessing because it has fueled a major program of economic and social development and—occasionally—enriched the people.

When Iraqis had to depend almost entirely on agriculture, they were few. When I first lived there in the 1950s, there were about 5 million Iraqis. The population has multiplied by five over the last half century despite war, sanctions, and repressions, to about 24 million today. About one in three live in Baghdad, Mosul, or Basra, and since about half the total population is today below fifteen years of age, increase, particularly in the cities, will be rapid.

In summary, for better or for worse, Iraq and the Iraqis are and will remain significant to the whole world economy, stability, and peace.

The third question is "*What* is distinctive about Iraq?"

That is, what makes Iraq unlike Mexico or France or Russia? This is an essential and recurring question for this essay.

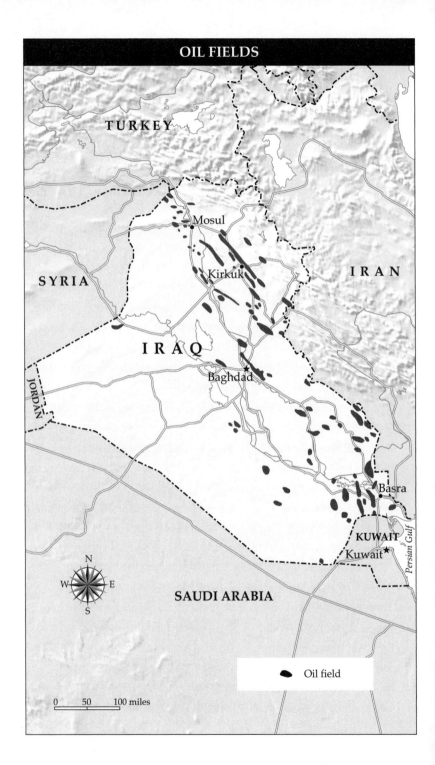

OIL FIELDS

The answer is to be found in its history. What is particularly distinctive are echoes from the earliest times of themes, attitudes, fears, hopes. Even when they do not "know" their history, Iraqis are guided by it. We who come from afar must listen closely to these echoes if we wish to understand modern Iraqis. It is for this among other reasons that this book aims to portray the long reach of Iraqi history.

That long reach is actually built into the land of Iraq. While there are fewer grand monuments like the pyramids of Egypt, because Iraqis built in mud brick rather than stone, the mud brick itself became a sort of living history. High above the flat plain of northern Iraq rise a number of mounds that look like long-extinct volcanoes. On top of them, modern towns have been built. The people who go about their daily chores in them are only vaguely aware that under their feet are not volcanoes or hills but hundreds of feet of rubble, a history of their ancestors in dust and stone. As each generation has lived and died, built houses and torn them down, brought in supplies and cast out refuse, their hill has grown. The people today do not know much about those who lived in the towns below their feet, but their lives are shaped by the contours of the hill and, to some unknowable extent, unconsciously, by memories of those who lived there before. In part, this book is an attempt to dig into those hills of memory to understand the foundations of the present.

To get what I believe to be the best view possible, I start at the very beginning. How the early peoples came to Iraq and what they did there to found "civilization" is the subject of chapter one. In chapter two, I deal with the coming of Islam, its modification of Iraqi society, and the heritage it has left behind. Chapter three takes up the formation of the modern state under

British overt and covert rule from 1914 to 1958. In chapter four, "Revolutionary Iraq," I show what happened when the monarchy was overthrown, British influence reduced, and authoritarian trends, already inherent, greatly magnified under a sequence of dictators ending with Saddam Husain. Chapter five covers the period of American domination from the Gulf War of 1991 to the partial turn back of government to an American-appointed regime. Finally, in chapter six, I bring together all these trends to set forth my estimates on the future of the country and on its relationship with the rest of the world.

ONE

═══════════════

ANCIENT IRAQ

Ancestors of inhabitants of today's Iraq began to emerge from the long shadows of prehistory about twelve thousand years ago. We cannot see them clearly, but we have some notion of how they lived. Gathered in groups of fifty or so people, they ranged along the slopes of the mountains that divide modern Iraq and Syria from Turkey. They did not live in permanent villages but sheltered under lean-tos that were covered with the skins of animals. The men hunted wild animals while the women gathered wild grasses from which they extracted the seed and pounded them into digestible bits.

Using crude sickles, faced with flint chips, they scoured the valleys to collect every edible thing, and, driven by hunger, they ate everything they found. Where archaeologists and paleo-botanists have made studies of their campsites, they have found more than a hundred different kinds of seed. Animals and seed were usually plentiful, but each day brought risk. Their besetting fear must have been famine. Although on average their lives were reasonably easy, they lived literally hand to mouth, so a chance break in the weather might send the wild animals on which they depended out of reach or cause wild grasses to wither.

As they hunted or gleaned one area bare, whole clans would pick up their few possessions and move to a new location. Often they left caches of seed behind in clay-lined baskets or pits to which they planned to return. It is astonishing, in these circumstances, that they left a rich legacy: they became our first farmers.

No one knows exactly how they began this revolutionary new venture, but the agricultural revolution was probably partly accidental. From time to time, someone, perhaps a child, spilled a basket or dropped a handful of seed. Probably also, at least some of the caches of seed they left in storage pits or clay-lined baskets got rained on. Much, of course, would have just rotted. But, over the long years, some would spout into what modern gardeners call "volunteers." Watching this, the tribesmen—and particularly the women—would naturally find it convenient that the volunteers were in or around their camps and water-holes where they were easy to collect.

At various times when and where rainfall was abundant in the hills along the northern reaches of what is now Iraq, some people began to winnow out seeds or sprouts. From the results, we know that what had probably been an accident certainly became a purposeful move. With pointed sticks like the ones they used to dig up tubers (and the men used to spear wild goats), they poked holes in the soft mud beside a waterhole or on the bank of a stream and dropped in a few grains. Probably the seed often failed to come up, but some did. The lucky or those who did it right were more apt to survive periodic famines than their more backward neighbors. Fear of starvation was a great teacher. Paleobotanists believe that within a few generations these adventurous—and hungry—tribal peoples achieved the first feats of domestication. So evident were the advantages of these experiments that the example spread widely from encampment to encampment. Sometime around 6000 B.C., "farming" began.

During these years also, the once-bountiful wild game had become harder to find and kill. Some of the little groups scattered along foothills and in the valleys of the Zagros—the area that because of its relatively bountiful rainfall has become known as the Fertile Crescent—had already begun a process that has been termed "management." Much as Laplanders today work with still-wild reindeer, they followed and partly managed herds of goats. Although goats are today reviled as destructive animals spreading ruin in fragile ecosystems, they offered great advantage to these early farmers. From accumulations of bones in their campsites, we know that goat meat was a big part of their diet. They killed enormous numbers, but some of the hunters must have figured out that killing more than they could eat just produced rotting carcasses that attracted dangerous carnivores. Since there was no advantage in killing all the animals, there might be a way to postpone the deaths of at least some of them. Keeping them was relatively easy since goats are extremely adaptable and thrive and breed with little forage. They were a major contribution to survival of those tribesmen who took the trouble, since in addition to meat, goats produce milk, which tribesmen began to drink; fiber for clothing; tallow for light and medicine; bone for tools; sinew for bindings; and dung for fuel.

Herding and then domestication of animals immediately created two new conditions that were of great and lasting influence. For the first time, little settlements had reasonably secure sources of a balanced diet. Fewer people died of hunger, and more lived longer. As populations increased, the initial trends were reinforced. Planting crops made staying put both possible and necessary, but managing or herding animals required some degree of movement. So divisions of labor, which had long been practiced, became more sharply defined. While the women, the elderly, and the very young remained fixed in the village, young men went out, sometimes for weeks or months at a time, to hunt or herd animals. This pattern was so stamped by generation after generation of daily experience that, until just a few years ago, it was the common way of life throughout the villages of Iraq.

Then, sometime about 6000 B.C., the mountains, foothills, and plains of Iraq became hotter and drier. Upland areas that had given rise to farming could no longer sustain the increasingly dense concentrations of people. To survive, many migrated to

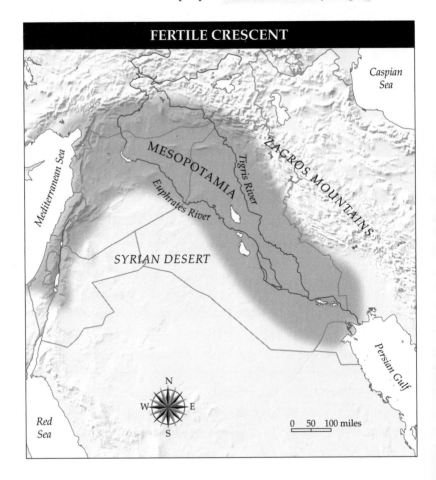

places where there was more water. Animals probably showed them the way in their yearly trek from the highlands in the summer down into the plains in winter. Following them, the hunters probably had already established temporary camps along the great rivers. As little societies grew beyond their resources or fell afoul of unsolvable domestic conflicts, their women and children would have followed them. Even if these moves were originally temporary, many people began to stay.

While much of the area of the plains was not as attractive as the mountains had been, it gradually became more so. The same techniques that had made agriculture possible in the cold north worked in the marsh and swamps of the warm south. Instead of rain, the rivers provided water in periodic floods, and some of the water was caught in the network of streams and ponds that then formed along the main channels. As word spread, the mountaineers flocked into the south. At first their lives were very hard, but after four dry and hot centuries, the climate changed again. As rains became more bountiful and regular, small groups of tribesmen who had become rudimentary farmers began to move farther south onto lands between the Tigris and Euphrates, where they settled and built scores of little villages. There, for hundreds of years, the people archaeologists have named the Ubaidians (after one of their sites) settled literally in the mud of the great rivers. There they began the second great revolution in agriculture, digging shallow ditches and dams to control the waters. Their more elaborate form of agriculture forced them to coin a word for "furrow"; to work it, they invented a revolutionary new tool, the ancestor of the plow. Growing richer from the bounty of the rivers, they began to turn the energies that had been mobilized for agriculture to other pursuits. We know this because later immigrants to the area, people who spoke a different language, borrowed their words

not only for furrow and plow but also for carpenter, weaver, pot-
ter, metalworker, and mason. Perhaps most impressive of all for
the future of Iraq, they had learned how to make bricks from
mud and straw. Thus they literally laid the foundation of Iraq.
In an area with poor resources of timber and virtually no stone,
brick made possible for the first time the building of permanent
villages. Some would become towns and a few would become
cities.

While the Ubaidian farmers toiled in their fields, another group
of people, perhaps from the same area in the north, began to
arrive in the south of Iraq. We call these people Sumerians from
the name of one of their settlements.

The Sumerians were one of history's most productive and
fascinating people. Indeed, with them, we can begin to speak of
"history." Much of what they did in a variety of fields laid the
social, economic, and religious bases for subsequent Iraqi civilizations. Yet, paradoxically, we can only speculate on who they
were. Our best clue is their language. Sumerian is not a member
of the family of Indo-European languages that includes English,
the Romance languages, Greek, Russian, Persian, and Hindi.
Nor is it a Semitic language like Arabic, Hebrew, or Akkadian. It
is a member of another family that linguists call "agglutinative,
semi-incorporating languages with vowel harmony." Sumerian
shares these characteristics with Turkish, Finnish, Hungarian,
Elamite, and the Dravidian family of languages of India. This
linguistic affinity suggests that speakers of an earlier common
language, probably nomads from south-central Asia, had fanned
out centuries before in a vast arc stretching from Finland, across
Europe through Iraq to India.

When it began to be recorded, around 3000 B.C., Sumerian

had already incorporated a number of words—including the names of villages—from Ubaidian. This suggests that the Sumerians went through a process similar to one that is better known elsewhere and later.* Incoming "barbarians" spend years, even generations or centuries, on the fringes of resident, more sophisticated societies, performing menial chores while they learn the culture and technology of the residents. This was characteristic of Germanic tribes along the frontier of the Roman Empire, Turkish tribes in the Byzantine Empire and Mongol tribes north of China. The fact that Sumerian contains words for crafts borrowed from the Ubaidians is evidence that the incoming Sumerians did the same in Iraq. Then, like the Germans, Turks, and Mongols, they became more intrusive and eventually took over the existing settlements. Many villages with Ubaidian names quickly grew into Sumerian towns.

Once living in towns, the Sumerians embarked upon a remarkable burst of creativity. Adopting the farming techniques of the Ubaidians, they elaborated them, rechanneling streams and ponds to create a more efficient and more widespread irrigation system. This soon began to produce a larger surplus of food, and that, in turn, promoted population growth. As towns increased in size around the fourth millennium B.C., the bonds of kinship and intimate neighborhood weakened and no longer sufficed to prevent destructive disputes. A few societies began to find substitutes for the pacifying mechanism of kinship. Among the earliest and most successful of these mechanisms was a religion-based patriotism. Virtually every town, or at least the ones

---

* For example in Greece, where the incoming "Greeks" similarly took the names for villages from the non-Greek natives whom ancient authors called Pelasgoi.

that survived, made itself into a cult center focused on a shrine or temple. Within its boundary stones, a god was enthroned as town's "owner." The town's fortune was explained by the favor or disfavor of the god. If a catastrophe occurred, everyone would know that the god had been displeased. The challenge was to know what the god wanted.

Some people claimed to have this special gift. How they convinced their fellow citizens we do not know, but we know that they did. These "interpreters" soon became a specialist profession to whom the community turned to supervise ceremonies designed to win divine favor. Acting in his name and managing his house, the temple, they assumed the right to demand gifts for the god, thus accumulating property that they "managed" on his behalf. Becoming rich, they became powerful. Being powerful, they were respected. In short, they were the harbingers of an autocratic style of government that has been a feature of Iraq down to the present.

While many peoples have created centers of pilgrimage, the continuity of sacred cities in Iraq is striking. Nippur, the temple of the god Enlil, was founded during Ubaidian times and was occupied for about five thousand years. While we can only speculate on how and to what extent such a long reach of the human experience impressed itself on the subliminal values and attitudes of a people, I think we must assume that it did. Somehow, in ways we do not understand, cultures are formed and survive over the tumultuous events of history. In Iraq, Nippur and other "holy" cities became unconscious prototypes for the later Shia holy cities—Karbala, Khadhamain, and Najaf. Like the Shia holy cities, Nippur was also a school where scribes and priests were trained.

Differences over the primacy of gods, as the emblems of towns, often justified hostilities with neighbors who followed other gods. Interurban wars became frequent. And these too set patterns that would echo down Iraqi history. In the course of fighting, as in other of the new crafts and trades, some men proved themselves more able than others. Such a warrior was known as a "big man" (Sumerian: *lugal*). The *lugal* was perhaps originally a landowner whose fieldhands constituted a readily available military force. Being called upon to defend his town enabled the *lugal* to enhance his wealth and power. So it was that by about 2800 B.C., what had been a personal attribute became an institution. This is what the Sumerians called *nam-lugal*, the "quality of being a great man," or, roughly, kingship. The cult of the "great man," once firmly fixed, has endured in the minds of Iraqis ever since.

Since, like ancient Greece, the Mesopotamian plain was divided into dozens of mutually hostile petty city-states, there was much scope for great men. They led their cities against their rivals, trying to encroach on them or seize their goods or people. Political and religious diversity thus unleashed on an unprecedented scale the force that was to shape Iraqi society—warfare.

In this warfare, the *lugal*'s retainers were necessary but not sufficient to defend the town. Enemies might come too quickly for a defensive force to be called in from the fields. So those towns that could afford the vast cost began to create protective shields against avaricious and dangerous enemies. The labor that had been mobilized to elaborate irrigation canals was soon turned to building walls. Most of these were small because the towns they were to protect were correspondingly small, but some became huge. The most impressive, built around the rising city of Uruk (biblical Erech) was ultimately nearly 6 miles, or 10 kilometers, in length. Parts of it were double and much or all of it was at least 23 feet, or 7 meters, high. The labor involved in piling up millions of bricks to form it was prodigious. It could be performed only by large societies. By about 2500 B.C., Uruk's population reached about fifty thousand people. Small towns could not compete, so over the following centuries, Uruk leached the settlements around it. As people flooded into the place where they would be safe, Uruk's 126 neighboring towns shrank to only 24.

Walls could protect against foreign enemies, but almost as dangerous as they were domestic rivalry, envy, and anger. As populations increased and wealth accumulated, it became as imperative to find ways to bring order into the distribution of goods among the inhabitants as to defend them from foreigners. The domestic counterpart of the wall was the deed. At first the deed was simply a set of symbols, but gradually after about 3000 B.C., the Sumerians began to use a kind of shorthand. Instead of trying to draw an accurate picture of an object, they pressed a triangular-shaped stylus into the clay to make more or less what we would call a stick figure. Gradually these stick figures became more abstract and stylized until they were formed into a pattern of writing known as cuneiform. Originally simple and concrete they became increasingly complex and abstract.

Both as the creators of writing and as its masters, a new class of scribes came into being. They were the first bureaucracy. We know about them both because some of their textbooks have survived and because, even then, people were complaining about their abuses and high-handedness. Already the Iraqi people had learned to stay as clear of administration as they possibly could. It was a lesson they would not forget under whatever central government ruled them for the next four thousand years. Even in the Iraq of the 1920s and 1930s, people tried not to sign documents or have dealings with government, even if ostensibly it was to their benefit. As we will see, for this Iraqis paid a terrible price when new systems of law were imposed upon them.

Because writing flourished there, we think of Iraq even in ancient times as an urban society. Cities were centers of religion, government, and trade, but they were fragile. Over the centuries, they rose, flourished, declined, and were abandoned. More enduring was the life of the peasant farmer. The Ubaidian would not have felt alien in the Iraq of 1900. Using the same tools, the modern Iraqi dug in the same soil, planted the same crops, ate the same food, and marched to the same rhythm of seasonal flood. For him, the grim reality of life was the climate. Southern Iraq has one of the harshest in the world. Under the intense solar radiation, plants and people literally wither. This seasonal reality was built into religion: the god Tammuz, like the Greek Adonis, was thought to "die" each year. Adonis descended into Hades, but Hades came up to Tammuz when the heat of summer became intense. Ancient Iraqis gathered to mourn his passing as modern Shiis gather to mourn the death of the *Imam* Husain. Unlike the Shiis who believe they will have to await Judgment Day to resolve the sadness they feel, the ancients believed that the goddess Ishtar descended to Hades to bring life-giving water as spring approached. For them, as for the ancient Egyptians, the cycle of life-death-life was a seasonal occurrence, god-ordained and god-regulated.

In Iraq, the poor were always tremendously exploited. They were forced to perform labor on canals and dikes, were drafted into military forces, and, above all, were made to stack mud bricks into walls and terraces and, eventually, into those most distinctive of Iraqi buildings, ziggurats. The demand for labor was insatiable. To build just one of the temple terraces of Uruk has been estimated to have required about five and a half million man-hours—that is a force of fifteen hundred men working ten hours a day for a year. Not even the inhabitants of a great city like Uruk could meet the demands. So, early in the development of the larger towns and cities, we find a new person coming on the social stage: the slave.

How slavery arose, no one knows, but my guess is that it was an extension of the process already well known for animals: domestication. The Sumerian word for "slave" is related to the word for "foreigner," which indicates that in Iraq as in Greece, many slaves were prisoners of war. It was easy to fit them into the urban environment as the city had already been segregated by class and function. Segregation thus became a mechanism of social control, but it was inefficient and insufficient because in the confines of the city, diverse peoples had to live together in perpetual neighborhood.

To maintain the existing distribution of property, the priest-hood and the monarchy, while often apparently vying for supremacy, were forced to work together. Together they not only organized the production of the city and kept order but also explained why the system existed. The urban elite believed that they were born into an international order; challenged and stimulated by other cities, they saw that order as reflecting a divine hegemony that both explained and justified the earthly experience. They evolved a complete system in which each city had its place; each vagary of human fortune, its explanation; and each person, from slave to ruler, his role. Iraq thus gave rise to the first systems of law.

The code of Hammurabi, king of Babylon from 1792 to 1750 B.C., is the most famous of these codes, but at least three others are known from even earlier times. Hammurabi thought that his code was so comprehensive that reference to it would answer all the questions of his subjects: they would find how and whom they should marry, what happened to property on death, how much interest was to be paid for various kinds of loans, and answers to a remarkable range of other topics in some 282 (surviving) sections. There was no need to add to the code and little need for interpreters of it. Having been accustomed generation after generation to rigid codes of conduct, later Iraqis would take easily to Islam, which similarly specifies answers to virtually all legal, social, criminal, and even culinary issues. The answers are embodied in the Quran. Like the ancient rulers, later Muslim jurisconsults regarded attempts to revise what was then proclaimed as the Word of God as unnecessary, an innovation, a heresy. The Muslim God, like the God of the Old Testament, is a remote and stern figure, best envisaged as a judge, who demands that his creatures live by an exact formula. That would not have seemed unusual to an Iraqi four thousand years ago. How ideas, tastes, fears, and customs linger is one of the unanswered questions of history; no one knows how they are perpetuated, but we know that they are.

In addition to the emphasis on "great men" and on rigid codes of conduct, consider the perpetual dream of the Garden of Eden where life was simple, pure, and satisfying. In its most concrete form, we know that gardens have been a distinctive feature of Middle Eastern culture for millennia. Given the rigors of the climate and the presence of that early acquisition of the people of the Zagros, the goat, gardens had to be protected from marauding animals and hungry humans. Gardens were the luxuries of the rich and powerful. So those who owned them thought of them as enclosed and protected sanctuaries. The Persians, probably borrowing from an earlier name for them, called them places "surrounded by a wall." Their word was *pairi-daeza*. Then, when Alexander the Great's Greek soldiers, undoubtedly hot, tired, and thirsty, saw them, they picked up the Persian word. For them, the gardens were truly *paradeisos*. And so the concept spread. Paradise is our heavenly reward. The early Muslims, coming from parched Arabia, who esteemed the lushness of a garden and the delight of running water as "heavenly," picked up the same notion, so that in the Quran, heaven is described as "a garden beneath which rivers flow."

Iraq not only influenced (and was influenced by) Islam, but several motifs of the Bible can be traced directly to it. The flood from which Noah was said to have saved our ancestors and the animals is an echo of an Iraqi myth: the god Enki warned a man named Utnapishtim to build himself a boat because the gods were angry with mankind and had decided to destroy it. Perhaps even more striking, Iraq gave us the ultimate Horatio Alger story—centuries before Moses, the baby Sargon was said to have been found floating in a little basket amid bulrushes on the great river Euphrates.

Meanwhile, another addition was being made to the Iraqi population. Just as the Sumerians infiltrated Ubaidian society, so Semitic-speaking peoples were migrating eastward from the Mediterranean across the relatively hospitable lands of what is now northern Syria. One group, whom we know as the Assyrians, stayed north, while others, whom we know as the Akkadians, followed the Euphrates south onto the Sumerian plain. The southerners began to congregate around the existing Sumerian towns, and as the Sumerians had done with the Ubaidians, they did menial work, learned, and eventually moved in. By the middle of the twenty-fourth century B.C., they found a leader, a man of such genius that even the dominant Sumerian civilization accorded him front rank in its historical records. He was Sargon I.

Sargon is the first statesman to emerge on the stage of history. Endowed with daring and brilliance, he both used the Sumerian system and remained alien to it. If we believe the legend, he took the remarkable step of building a political base in the only way then understood: he built the town of Akkad to house his Semitic-speaking followers, the Akkadians. They were to supply a coherent, self-supporting political core for the empire he began to construct. His reign would not be the last time that the Iraqis were prepared to give up their freedom to a strong leader who promised them security and prosperity, and in our times, we shall see that Saddam Husain unconsciously followed Sargon in focusing his rule on an inner core of related followers. Supported by this inner core, Sargon attacked and, one after another, destroyed the walls of the old Sumerian cities. In thirty-four battles, he beat down his opponents, also as Saddam Husain would later do, and unified southern Iraq. Then he moved north to secure the sources of raw materials in Syria, parts of Anatolia, and what is today Kurdistan. His outward thrusts, the first great imperial conquests of history, were spectacular but costly. As they chafed under new impositions and came to resent Sargon's officials, some of the cities revolted. Again, as Saddam Husain would do, he savagely suppressed them. But unlike Saddam Husain, he was able to pass his state on to his heirs. His grandson Naram-Sin, who ruled from about 2254 to 2218 B.C., carried to its full potential the new concept of a unified Iraq under an imperial leader.

As an imperial leader, Naram-Sin set a style that would be copied by ruler after ruler down to our times: he took a grandiose title. His was "King of Totality." Centuries later, Sargon II would proclaim himself "King of the World." And in his turn, upon becoming king of Babylonia, Cyrus took the same title in 539 B.C. In our times, Abdul Karim Qasim proclaimed himself the "Sole Leader" and Saddam Husain liked to be called "Hero President." Iraqis evidently have had a deep-seated desire to be—and to venerate—"great men" since they first coined the term *lugal*.

Sumerian, like Latin in the medieval West, would soon be overlaid by a form of Semitic. The amalgam of Sumerian and Akkadian, what we know as the Babylonian civilization, would remain in the minds of men, century after century, as the embodiment of the cultural history of Iraq, occupying the exalted position later Europeans accorded to the classical age of Greece.

While this rich tapestry was unfolding on the plains of the south, another society was gaining power in the northern hills. The original Assyrians we have already met as a branch of the Semites who had moved into Iraq along the fringe of the Fertile Crescent from what is today Syria. Like their cousins who had gone south, they had been strongly influenced by Sumerian civilization and had formed a small city-state near modern Mosul. Like other Iraqis, they were known to their neighbors mainly as farmers and merchants. Then, from about 1350 B.C., they underwent a major transformation into a military state. Formidable though they showed themselves to be, they could not match the power of the great empire of the Hittites. So it was not until a period of chaos in the tenth century B.C. that the way was open for them to achieve their military potential. That achievement was one of the great military explosions of all time.

Assyria occupied only about 5,000 square miles, or 12,950 square kilometers (about the size of Connecticut); its climate was harsh, and local resources were meager. Like Alexander the Great's Macedonia, it had a small native population; probably it was no more than one hundred thousand. Also like Alexander's Macedonia, Assyria became an engine of war. Its rulers proclaimed that war was the natural condition; it was just; it was god-ordained; Assyria was its earthly embodiment. Lesser peoples must submit. The Assyrian term for submission was "walking on all fours" (*eli erbi ritti pasalu*), that is, to become like a domesticated animal. Those aliens who refused their proper place in the Assyrian world order must be eliminated, their towns razed, and even their gods carried off.

Massacres and enslavements were celebrated in massive wall paintings and sculptures with which the kings decorated their palaces and city buildings. Scene after terrifying scene shows the vanquished being mutilated or killed, villages being sacked, battlefields littered with the bodies of the fallen, and prisoners working under the watchful eyes of soldiers. Propaganda as well as superb military technique was fully exploited to make up for the small size of the armies. In 889 B.C. Tukulti-Ninurta II began what would be a half century of unprecedented conquests. He and his immediate successors must rank as some of the greatest— and least known—generals and administrators of all times. They were driven by a compulsion to triumph that became virtually a national religion, as different from the loyalties and powers of the old city-states as their "steeled" iron weapons were from the baked clay weapons of the early Sumerians.

The later Assyrian rulers retained the heritage of their merchant ancestors. When they conquered a town or seized a province, they made and proclaimed publicly an inventory of the spoils. They sought, however, not only spoils but, as one of their emperors put it, "the elimination of local characteristics." In part to accomplish this, the Assyrians are thought to have relocated as many as four or five million people from parts of their empire to others. This vast whirlpool of peoples brought about, among other effects, the homogenization of the peoples of Iraq. Theirs was a pattern that, unconscious of history, Saddam Husain would follow by moving Arabs into Kurdistan and Kurds into the Shia provinces of Iraq in our own times.

Militarily brilliant though their rule was, the Assyrians

inevitably exhausted their resources. As their already small population fattened on the spoils of conquest, the state began to employ large numbers of foreign mercenaries from the surrounding mountains. What happened cannot be documented, but it must have played a role in the creation of the forerunner of a "Kurdistan." As the state came to depend increasingly on foreigners, it gave up its original exclusiveness and began to incorporate them. They, like the Sumerians, the Semites and others, soon learned the "secrets" of power and were ready, when opportunity came, to use them against the Assyrians, who had lost the vigor of the rise to power and the mystique by which they held it. The end came suddenly. Assyria's riches and its reputation became more targets than shields; by 616 B.C. it was revealed as just a small power among others. Assyria, state within an army, army as a state, defeated on the field of battle, would never rise again.

Astonishingly, it was the remnants of the old Sumerian city-states, now focused on Babylon, that administered the final blow to Assyria. Having been destroyed by the Assyrians in 689 B.C., Babylon was rebuilt and reached a new apogee around 600 B.C. under Nebuchadnezzar II. Nebuchadnezzar is best remembered today for having built his hanging gardens—allegedly to soothe the longings of his wife for her home in the mountains of Iran—and for putting down rebellions against his rule by the kingdom of Judah. Having taken Jerusalem in 597 B.C., he had to put down further rebellions over the next twenty years. Following the Assyrian pattern, he deported group after group of Jews to Iraq. Two thousand years later, some Iraqi Jews believed that they were descended from those whom Nebuchadnezzar had brought there.

It was not only to Palestine that Nebuchadnezzar advanced. He created an ephemeral empire that stretched into Egypt, along the Mediterranean coast, and into Iran. But small city-states were too small to sustain great empires in the turbulent world the Assyrians had left behind. So, although Babylon went through a brilliant reinvention of the old Sumerian-Akkadian culture, its days were numbered. We know more about them than about many of the earlier, and more important, periods because they figure in the Old Testament. But power had shifted to the east: the combined peoples known as the Medes and the Persians had formed a new state that would become in its time the world's greatest empire.

In 539 B.C. the new Persian emperor, Cyrus, defeated the Babylonian army and entered Babylon. At that time Babylon was a flourishing city whose merchant house of Egibi Sons were the Rothschilds of their times. So attractive did Cyrus and his successors find Babylon that they made it the administrative capital of the empire. More important, Cyrus attempted to bring about a synthesis of Persian and Iraqi culture. The quest for this synthesis laid the foundation for the great dilemma Iraqis face today.

One result of this quest was the Persian recasting of the ancient Iraqi tradition of gathering to recite or listen to "tellers." Precursors of these men had earlier narrated what we know as the Babylonian Epic of Creation. In later but still ancient Iran, reciters repeated the national epic, *The Shahnameh*, and sang "The Weeping of the Magi" (the Zoroastrian priests). After the coming of Islam, the text changed to one appropriate to the new religion, but the form continued: reciters (Persian: *rowzeh-khans*) narrated the martyrdom of the *imam* Husain. Their counterparts elsewhere, the *rhapsode* in ancient Greece who recited Homer, the *bard* who sang Celtic epics, or the *soɣgu-maδ* in

Iceland and Norway who recited the sagas, shared the task of reinforcing cultural identity and keeping alive the traditions of their people. In Iraq, the reinforcing of tradition was carried to a relatively massive scale: Persian Zoroastrian priests, known as magi, were established in their own town near Nippur. Perhaps it does not stretch the evidence too far to suggest that they formed the model for the later Shia clergy, who similarly had their own towns in both Iran and Iraq and who were considered the guardians of sacred knowledge. It is, in part, in this way that the remarkable continuity of Iraqi culture, which is a theme of this book, was kept alive.

In more mundane matters as well, the Persians influenced Iraq far into the future. Perhaps most important, they laid the foundation for the land ownership system that, essentially adopted by the incoming Muslim invaders, set the basic pattern for the life of most Iraqis for the next two thousand years. Finally, while I have used the term "Iraq" throughout this essay, it is appropriate to acknowledge that it was the Persians who coined the name. The Arabic looking word Iraq (Arabic: *al-'Irāq*) actually comes from the Persian *erāgh*, which means simply "the lowlands."

While the Persians enjoyed Iraq, they craved Greece. And having watched the Iraqi cities hack one another to pieces and so make their conquest easy, Darius and Xerxes tried to apply the Iraqi lesson to Greece. In one of the great turning points of history, they failed, but they opened the way for one of their allies to succeed. Macedonia, then ruled by King Philip, was a western version of Assyria, a state whose industry was war. When the Persians withdrew, Philip's Macedonians were ready to attack other Greek cities. On the brink of success in his Greek strategy, Philip was assassinated. Not to be cheated of victory, his army chose his son Alexander as his successor.

For the remaining eleven years of his life, Alexander would attempt to conquer the vast Persian Empire. In 334 B.C. he crossed into Asia. The Persian ruler sued for peace, but Alexander wanted glory. So he threw his disciplined small force into the Persian host at Gaugamela near the modern Iraqi city of Irbil. From that victory, Alexander moved into Egypt; then doubling back, he entered Babylon in 330 B.C. From Babylon he marched across Iran into Afghanistan and on to the Indus. There he faced a revolt and began the return march toward Greece. When he arrived in Iraq, he conducted what must be the most bizarre ceremony of all time: to symbolize the meeting of East and West, as he understood them, he arranged that all his soldiers would marry their camp followers in "the Wedding of the Ten Thousand" while he married a daughter of the Persian king. His plan, symbolized by this event, was the same as Cyrus's—to make Babylon the capital of the world.

When he died shortly thereafter, Alexander left no heir. His army split apart, with each commander grabbing a piece of his empire. Seleucus took for himself the satrapy, or province, of Babylonia. Not excited by the mysticism of Alexander's last days, he regarded Babylonia as the springboard to world empire. But, finding it too poor and depleted by war for his needs, he soon moved away.

Meanwhile, far to the east beyond the world the Macedonians knew, a major new force was coalescing from the tribes of Central Asia. It would become known in history as the Parthian Empire and was brought together during the second century B.C. The Parthians took Babylon in 144 B.C. and at a site on the Tigris, south of modern Baghdad, called in Persian Tespon, which we know in its Greek guise as Ctesiphon, they built their capital. Its massive audience hall, much of which is still standing, soaring up 121 feet, or 37 meters, is the world's highest brick arch still standing. It must have overawed visitors, particularly nomads from the desert, who increasingly were encroaching on the settled lands.

The Parthians treated the Euphrates as their western frontier and periodically fought the Romans in Iraq. In their most famous encounter, the great battle of Carrhae (Harran) in northern Iraq in 53 B.C., their cavalry wiped out the slow-moving force of the Roman general Crassus. Carrhae was Rome's greatest defeat. Warfare swayed back and forth across Iraq. Then, in A.D. 224, they gave way to another group of Persian invaders who established the Sasanian Empire. Iraq was to become the center of their domain. Like the Parthians and the Romans, the Sasanians were locked in combat almost continuously for generations with the Eastern Romans, or Byzantines. With their eyes fixed firmly on each other, they allowed the defense of their southern frontiers to fall in disrepair. Then, in 570, far beyond the world known to them, Muhammad, the prophet of Islam, was born. His life would open a new era in the history of Iraq.

TWO

ISLAMIC IRAQ

Far away to the southeast from Iraq, in the oasis trading towns of Mecca and Medina, a new religion was being born. When Islam reached Iraq, it would become interwoven with existing institutions and ideas, transforming both but being transformed itself. Understanding Iraqi history and current events without an appreciation of what it was and what it became is impossible. Here I will single out those elements that are essential to Iraqi history. I begin with Islam's birthplace, Mecca.

Viewed from Europe, Mecca was almost unbelievably remote. Going there from France or Italy would have been an arduous and dangerous expedition of many months by sailing boat and camel caravan. However, Mecca was not quite as provincial a town as this suggests. Viewed in its own perspective, Mecca was the focal point of a complex of trading routes. Caravans regularly went back and forth to Yemen, where Meccan merchants met traders from India and the Spice Islands. Other caravans crossed the Great Nafud desert to the junction of the Tigris and Euphrates rivers at the Persian Gulf where they met Persian and Central Asian merchants. Some went north to

Damascus, where they bartered with traders from all parts of the vast Byzantine Empire and from the new towns of southern Europe. Traffic went both ways. Mecca was at least visited by groups of Christians and Jews. And Arabs came from all over the Arabian peninsula to attend yearly fairs, enter into poetry contests, and worship at Mecca's shrine, the Kaaba. Judged by the standards of the time and place, Mecca was a cosmopolitan center.

Mecca was also crassly materialistic. It was a merchant community, heedless of its poor and downtrodden, with its eyes firmly fixed on commerce. Until he was about twenty-five years of age, Muhammad was a merchant and presumably shared the values of his colleagues. Then he began to find the attitude of his fellow citizens deeply disturbing. Occasionally, he left Mecca for the desert, to fast and reflect on the sins of mankind.

At this point, the non-Muslim historian pauses to look at Muhammad's personal experiences and motivations—his knowledge of the world, his alienation from his community, and his mysticism. To a believing Muslim, these personal attributes appear irrelevant: what mattered was that in his unknowable wisdom, God visited Muhammad with his commandments for mankind. Muslim and non-Muslim agree, however, that when he was about forty, Muhammad embarked on his religious mission. As he told his kinsmen and friends, and as the Quran (Koran) confirms, he had a vision of the angel Gabriel, who ordered him to "Recite in the Name of the Lord." The stunned and frightened Muhammad is said to have stammered, "But, what shall I recite?" Getting no immediate answer, he recounts that he spent a lonely and frustrating period without further guidance. During this time, he often went apart alone in the desert, meditating and fasting. Finally his visions resumed. Then he began to transmit to those who would listen, in a stream of commandments that poured forth until his death in 632, what his followers much later gathered as the Quran.

Muhammad took no personal credit for these commandments: he described himself as merely a messenger (Arabic: *rasul*) for God's Word. Other men, his predecessors, he said, were true prophets. They included not only virtuous folk figures of Arabia, but Jews from the Old Testament and, above all, Jesus. Jesus, we learn from the Quran, towers above Muhammad in God's favor. While the Quran denies that Jesus was the son of God "who neither begets nor was begotten," it also proclaims that Jesus was so close to God that he alone of all men was allowed to perform a miracle. Islam makes no such claim for Muhammad. What Muhammad says he did was only to bring God's message, the same message that had been delivered earlier to Moses and to Jesus, to the Arabs in their language, Arabic.

The Meccan oligarchs were furious. Mecca was the center for cults that justified their position in the city and were woven into their commerce throughout Arabia. In their eyes, Muhammad's message was treason. They decided to kill him. That was dangerous while his close relatives protected him. So the city leaders put pressure on them to repudiate him. When they appeared about to do so, an act that would have made him, by local custom, an outlaw, he fled the city.

Some months later, inhabitants of the little town later known as "the City [of the Prophet]," Medina, invited Muhammad to arbitrate a long-standing local dispute. That invitation gave him and the little group of Meccans he had convinced of his mission a new opportunity. Wisely, Muhammad sent his followers ahead so that when he arrived, it was as a prophet armed.

The town he encountered was far more primitive than Mecca. It resembled the little agricultural and animal-herding towns of Iraq during the Ubaid period several thousand years before. Indeed, it was hardly a town at all. More a scattered collection of hamlets, it had no urban institutions. It was divided, occasionally bitterly, into several clans of Arabs and a small community of Jews.

Muhammad's first task was to integrate his followers. To do this, he arranged for each immigrant to be adopted as a "brother" by some member of the resident Arab groups. Like most things Muhammad did, this was to set a precedent for the future: all Muslims are enjoined to be brothers (Arabic: *ikhwan*) one to another. On this basis, he made peace among the Arab clans and the resident Jewish community in an arrangement that has rather grandly been called "the Constitution of Medina." Both the emphasis on law and the recognition of a protected position for non-Muslims would also indelibly mark Islamic society.

Muhammad's second task was to protect the town from marauding bedouin. The bedouin, whom as an urban man he profoundly distrusted, were in the pay of the Meccans (who used them to guard caravans) and were accustomed to raiding agricultural lands when they ran short of supplies. Against them in the aggregate, Muhammad had no military capacity. But he knew that they were at war among themselves. This fact not only saved his community then but profoundly influenced Iraqi history up to the present time. It is a part of Arabic history that is not well understood. Briefly, it can be seen as follows.

Because of the limited resources of the desert, no group could be large. The "tribe" of hundreds or thousands was only a theoretical unit. In practice no group larger than fifty or so individuals could stay together because their animals would exhaust the nearby grazing and water. The effective unit, that is, the group that tented together, herded animals in common, and protected one another, was usually made up only of the descendants of a single man over a few generations. Within this "clan" (Arabic: *qawm*) there could be no fighting, while among clans there could be no permanent peace.

What Muhammad did was to recast the incipient Muslim community into the form of a clan. As in a kinship-based clan, in his new religion-based clan, there could be no fighting; "brothers" had the obligation to protect one another. But, unlike a traditional clan, which had no way to align itself with other clans, the new society welcomed adherents provided they professed Islam. The new society, the "religious clan," soon became larger than any kinship clan. What happened then was almost mechanical: the entire force of the new Islamic society was brought to bear on single nomad clans, one after another. Each clan found itself unable to resist its traditional rivals, other bedouin clans, and the Muslims. Since it knew of no customary way to amalgamate with traditional rivals, it risked being crushed between them and the new Muslim society. The only way it could protect itself was by joining the only side that was prepared to welcome it: the Muslims. Each new addition made the Muslims even more overwhelmingly powerful. So, like a desert sand storm, Muhammad's followers swept across Arabia. When he died in 632, eleven years after he had fled Mecca, almost all Arabia belonged to his "clan."

Spectacularly successful as the rise to power had been, its collapse was likely to be even more rapid. The bedouin looked upon Muhammad as they looked upon clan patriarchs. Loyalty was owed to him, not to the new religion. As the Quran chastises the bedouin, they were merely those who "submit" (Arabic: *muslim*) not those who really "believe" (Arabic: *mumin*). Although Muhammad must have realized how fragile his community was, he did not plan for what he must have realized would happen upon his death. For their part, his inner group of followers apparently so venerated him that they were shocked, almost numbed, by it. They had no precedents upon which to draw.

What happened then was partly accidental. Muhammad's closest associates knew that he had been planning a foray northward into Byzantine territory. They decided that they should honor his plan. The raid was to go ahead. It did and was successful. When the raiding party returned, they found nearby bedouin on the point of attacking Medina. We do not know exactly what happened then, but apparently the bedouin were so astonished by the booty the raiders had brought back that they quickly rediscovered their loyalty.

The implications were not, of course, lost on Muhammad's followers. They recognized that they must find a leader to continue at least the civil aspects of Muhammad's activities. To this end, they selected the man who had occasionally led the daily prayers when Muhammad was indisposed. He was the person who "stood in front" (Arabic: *imam*). So this man, Muhammad's father-in-law, Abu Bakr, was named as his "successor" or caliph (Arabic: *khalif*). Abu Bakr immediately decided to repeat the gambit that had saved Medina, a raid on the rich lands of the north.

Raiding the rich lands of the north was traditional among the hungry, threadbare bedouin of the Arabian deserts. To stop their raiding completely was impossible. The bedouin were highly mobile: on their camels, they could attack, plunder, and retreat into the desert before the cumbersome infantry of the Byzantine Romans or Sasanian Persians could muster defenders. So both of the great empires had opted for a sort of patrol (Arabic: *badia*) composed of Arabs to police the desert frontier. The Byzantines recognized and subsidized a group known for their leaders as the Ghassanids that ranged along the settled boundary in what is now

Syria and Jordan, while the Persians did the same with a group known as the Lakhmids who headquartered at the little town of Hira in Iraq. This arrangement was cheaper and more effective than stationing garrisons everywhere the bedouin might strike.

The system worked well for generations, but in the early years of the seventh century, the two empires had fought each another to an exhausted standstill. In 611 the Persians had invaded Byzantine territory, and in 614 captured Jerusalem; the Byzantines had rallied and fought back. Doing so disrupted tax collection, destroyed crops, and displaced or killed large numbers of people. Both empires were so short of money that they cut the subsidies they had paid to their Arab "policemen." The Persians dispensed with the Lakhmids altogether and appointed a Persian as governor of Hira. That was shortsighted at best, but what the Byzantines did was worse. They had long tolerated the sect of Christianity, Monophysitism, practiced by their Ghassanid clients, but in a sharp reversal, Byzantine authorities tried to force upon them Greek Orthodoxy. Neither the Byzantine Arabs nor the Persian Arabs were willing to defend their old patrons. What had been planned as a barrier became a bridge.

After a series of probes had revealed these weaknesses, the Arab invasions began in earnest in 633. In a lightning raid, using tactics the bedouin were good at, surprise and mobility, they rode to the outskirts of the Persian capital, Ctesiphon, just outside modern Baghdad, and then made straight across the desert to Damascus. Astonishing the Byzantine garrison, they looted the city. Then, hovering around Byzantine forces, the Arabs defeated them piecemeal. Combats, retreats, raids, and sieges went on month after month. While the Byzantine forces were depleted by the constant marching and fighting, the Arab forces grew with each encounter because bedouin were being lured from Arabia by tales of vast riches ripe for the taking. The climactic battle took place in the middle of the summer of 636 when the Arabs destroyed a Byzantine army led by the emperor himself on the Yarmuk River in what is today Jordan.

Probably as astonished as the emperor by the event, the caliph came to Jerusalem, whence Muhammad had dreamed he had ascended to heaven, to organize the conquests. Abu Bakr's only precedent was what Muhammad had done in Medina, and he essentially copied that model. Submitting Arabs were to constitute the Muslim community while non-Muslims—both Jews and Christians, "People of the Book [the Bible]"—were to live in peace, managing their own affairs and following their own religions, under the protection of the Muslims. No attempt was made later there or in Iraq to proselytize. As the Quran had said, Islam was the religion of the Arabs that Muhammad had brought to them in their language, Arabic. In any event, they were too busy conquering the world to spend much time on religion. Turning east, Arab tribal forces defeated a Persian army and captured the Sasanian Persian capital at Ctesiphon. Persian rule in Iraq collapsed.

In 644 a Persian prisoner took revenge. He assassinated the second of Muhammad's successors, the caliph Umar. Again the inner circle of Muhammad's followers met and selected as Umar's successor a weak and aged man who was quickly co-opted by his kinsmen, Muhammad's old enemies, the Umayyad clan, who constituted the Meccan oligarchy. Proving that blood is thicker than faith, the caliph Uthman began to distribute much of the booty being gathered by the victorious Arab armies as well as key governorships of provinces to his kinsmen. Muhammad's followers were outraged. Within a few years, rebellions had to be suppressed in Iraq and elsewhere. Finally, in 655, a group of discontented Arabs assassinated Uthman.

This group may be taken as representing the generation that had grown up in twenty years since Muhammad's death.

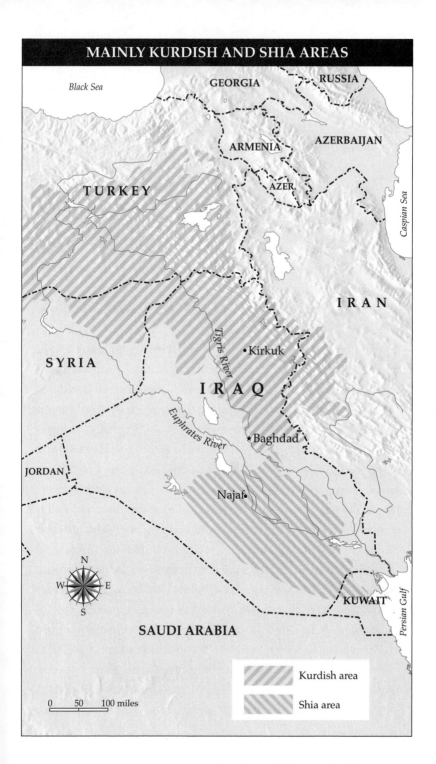

MAINLY KURDISH AND SHIA AREAS

Black Sea

GEORGIA

RUSSIA

ARMENIA

AZERBAIJAN

TURKEY

AZER.

IRAN

Caspian Sea

SYRIA

Tigris River

•Kirkuk

I R A Q

Euphrates River

★Baghdad

JORDAN

Najaf•

N

W   E

S

KUWAIT

Persian Gulf

SAUDI ARABIA

Kurdish area

Shia area

0   50   100 miles

Probably much of their discontent arose from their feeling that others had profited more than they. But there was a more significant reason. The nature of the Islamic "clan" had changed. Grown far beyond its original scope, it had become a community (Arabic: *ummah*).

The basis of new community, as Muhammad and his successors envisaged it, of course, was Islam. Joining was made easy. But the assumption of Muhammad and his first followers was that those who joined would be Arabs. When they said "Muslim," they also meant "Arab." Muslims should not seek to convert non-Arabs but should encourage them to become *like* the Muslims in their own communities. That was the basis of toleration of Jews and Christians.

Toleration certainly did not satisfy those who saw that the main beneficiaries of the new order were Arabs. To join the dominant community became a major aim of the conquered peoples. Some of those who converted were the children of Arab fathers and alien mothers, while others were Greeks, Persians, and peoples of other ethnic groups who were beginning to speak Arabic and follow Arab customs. The traditional Arab way for individuals to join a society was to become its "clients." For this status, the Arabs had a number of terms. The one that became widely used in the time of Uthman and later was *mawali*. Although their position could be explained in a traditional manner, the *mawali* actually occupied an unprecedented position: as converts to Islam, they were "accepted" as fellow Muslims, but, as non-Arabs, they were clearly not a part of the ruling elite. Those with the greatest skills and with access to important information, such as Persian bureaucrats, were often confirmed in their positions with associated privileges. But the vast majority did not have these attributes. They were relegated to a secondary status.

For those of the "marginal" people who were the offspring

of the Arab invaders and Iraqi women, discrimination was particularly obnoxious. Some were men of high culture who resented illiterate bedouin waxing fat off what had been the land of their maternal ancestors. Their resentment would become a major force in future events. Residues linger even today. It was in this sullen atmosphere that the cousin and son-in-law of Muhammad tried to establish his caliphate.

In the context of his times, Ali was a moderate. He was opposed by three parties, fundamentalists who became known as the "those who went out" (Arabic: *Kharijis*), a few rivals from the Muhammad's early followers, and the Umayyad clan who had profited from Uthman's rule. With relative ease, Ali overcame the old guard. The fundamentalists were a more serious threat. They had initially supported him but demanded that he condemn Uthman as a "tyrant" and so justify his assassination. When he would not do exactly as they demanded, they deserted his army. Without them, his army was no match for the Umayyad clan. Its leading member had been made governor of Damascus by Uthman. Muawiyah was a skillful administrator and had used his time as governor well; he alone commanded a well-organized army.

The major scene of action was in Iraq, where Ali had gone to try to recruit the Arab tribes who had settled there. Ali needed an army, but he realized that a crushing victory, even if possible, would destroy the very objective he sought. So he tried to use a shifting combination of fighting and negotiation, and, as is often the fate of moderates, he was misunderstood and opposed by both factions. After a series of pitched battles, truces, and conferences, he was murdered by a fellow Muslim in the new capital of Iraq, Kufa, in 661.

•   •   •

Ali's death cleared the way for the Umayyad clan, and they seized their opportunity. Drawing on the model of the Byzantine client kings of the Ghassanids, they created a vast new "Arab Kingdom," in which they legitimized and popularized their rule by a series of brilliant military campaigns across Africa, southern Europe, and Central Asia. Outwardly, the Umayyad caliphate achieved a military success rarely matched by any empire. Inwardly, the story was quite otherwise. The discontents that had begun in the generation after the death of Muhammad spread and grew more bitter. The conquered peoples, particularly those of Iraq, were attracted to Islam but were repelled by the Arab government. They expressed this in two ways: on the one hand, they asserted strongly their belief in Islam, but an Islam they defined in a non-Arab fashion, and on the other hand, they became involved in revolutionary movements.

Their opposition to the state focused on the caliphate. After the murder of Ali, Muawiyah held power virtually unopposed and arranged that upon his death, in 680, his son Yazid would succeed him. Since Yazid had no relationship to Muhammad, his followers, or his faith, revolts broke out in Iraq. There Arab tribesmen and their half-breed progeny invited Ali's son Husain, a grandson of Muhammad, to come to Kufa, where they promised to support him. He got little support and was soon besieged by troops supporting Yazid. When he refused to surrender, he was killed. The date of his death, the tenth day of the lunar month Muharram in the Hijrah year 61 (the Islamic year dates from Muhammad's flight from Mecca), has become the most important date in the calendar of the "Partisans of Ali" (Arabic: *Shiis*) ever since. For them, it is a day of unremitting shame: the time when they failed to support the man in whose body resided "the spirit of God."

Husain's death did not stop the revolt against Yazid; sieges of Medina and Mecca ended in the execution of many of Muhammad's closest companions. With them out of the way, Yazid was able to solidify the Umayyad dynasty. It was to last almost another century. But it was a century of periodic revolt and clandestine revolutionary activity in which the most dedicated were the "partisans" of Ali. Among them, the "martyrdom" of Husain unleashed emotional forces that were to have profound and lasting effects. Husain became the poignant figurehead of the revolutionary movement that would overthrow the Umayyad caliphate, the Hashimiyyah.

The Hashimiyyah was the most extreme of a number of anti-Umayyad, mainly Shia, movements that grew up primarily in southern Iraq in the early years of the eighth century. From Kufa the movement spread over the entire eastern part of the Umayyad empire among both Arabs and *mawali* who had converted to Islam but who were still strongly influenced by the Zoroastrianism or Buddhism of their ancestors. Through them, mystical religious ideas and practices were woven into the revolutionary movement. The very mystery of the movement lent it an emotional appeal that was lacking in rival movements. People flocked under its black banners from all over Persia and Iraq. By 747 it had become unstoppable. Town after town fell to the advancing rebels on the roads to Iraq. In 749 they crossed the Euphrates to advance on Kufa, where their movement had begun.

Then something peculiar happened. On the road to Kufa, the revolutionary movement was hijacked. While its adherents had marched to restore the Party of Ali to the caliphate, suddenly that objective was replaced by the aim of promoting a different branch of the Muhammad's kinsmen. How this happened, we do not know, but we do know that those who had fought for the Shia cause felt they had been cheated of the fruits of their victory. That would become a persistent theme in Shia experience over the centuries that followed.

The winners were the branch of Muhammad's family known from one of their ancestors as the Abbasids. The change was much more significant than merely the substitution of one branch of the clan for another: no sooner had the first leader of the Abbasids taken power than he revealed that he was not a Shii but a Sunni. Lest this seem merely a recondite matter, we should compare what happened in Iraq to a similar struggle between Catholics and Protestants in England: convulsed by religious conflict, the Protestant winners beheaded King Charles I and put Oliver Cromwell in power. When the monarchy was restored, Charles II and James II secretly espoused Catholicism. Their action, in turn, provoked the Protestant "Glorious Revolution." In seventeenth-century England as in eighth-century Iraq, politics was defined by religiosity so that the personal faith of the ruler became supremely important, even vital, to his followers and opponents. Both quickly learned that truth.

Having proclaimed his true belief, the first Abbasid caliph suppressed the Hashimiyyah and other Shia movements in Iraq while his armies drove the Umayyads out of Iraq, Syria, and Egypt. He had accomplished these tasks when he died four years after seizing Kufa. It was his brother, Mansur, who then in 754 began to organize the regime. During the quarter century of his reign, he created a new administration patterned on the old Sasanian Persian system and headed by a family recently converted from Buddhism, the Barmakids. Despite, or perhaps in part because of, the revolutionary past, Mansur turned the Abbasids away from the relatively open, Arab style of the Umayyads toward the more monarchical court ritual of the Persians and,

despite his emphasis on Sunni orthodoxy, favored Persian converts (Arabic: *mawali*) for government office. To house his
regime, Mansur built a new imperial capital, which he named
"the City of Peace" (Arabic: *Madinatu'l-Salam*) but the old name,
Baghdad, stuck and has been used ever since. It wasn't only the
name that proved enduring: the city was built atop ruins dating
from Babylonian times and the area in which I lived in the 1950s,
known as Bustan al-Khass, dates from Mansur's era. Through
fire and flood, invasion and plague, the city endured. It is one of
the most striking features of the Iraqis that, even when they forget their past, they preserve or recreate it.

Founded as a palace complex, Baghdad soon developed into
a huge and flourishing manufacturing and commercial center.
Like the ancient Sumerian cities and like medieval cities in the
Middle East, China, and Europe, it was divided by trade and
profession so that leather workers were in one area, carpenters
in another, and so on. Each of these districts was really a town
within the city; several were enclosed in separate walls, served by
separate mosques, schools, baths, and markets. In their districts,
the protected "People of the Book," Jews and Christians, were
governed by their own officials, collected their own taxes, and
worshipped in their own synagogues or churches. While their
lives were often not easy, the minorities lived a far freer and certainly a safer life than contemporary Jews or Muslims in Europe.
Even Zoroastrians, remnants of the old Persian religion who
were not "People of the Book" and so not officially tolerated,
also still lived in Baghdad. They were not many, but their numbers perhaps give a wrong impression of their influence. Even
among orthodox Muslims, Zoroastrian customs remained popular. The Persian feasts of Nō Rōz in the spring and Sada in the
fall were widely celebrated. Even goods that were deplored or
banned in formal Islam were openly sold and actively sought.

In some fields, Baghdad became preeminent. During the Dark Ages when few Europeans could read or write, it was famous for its manufacture of what was then a rare and esteemed luxury item, fine-quality paper. Some was shipped to still-literate Byzantium, but most was consumed locally in its thriving book trade. Far more literary than the city I knew in the 1950s, Baghdad then had more than a hundred booksellers at the Suq al-Warraqin (the paper sellers' market). Much of what is today known of early Arabic literature and even of the grammar and syntax of the Arabic language dates from that period and was produced by scholars of whom some were Persian.

Merchants and craftsmen were organized, as in contemporary Europe, in guilds. The guilds were supposed to maintain the quality (and honesty) of their trades, to ensure that taxes were paid, and to care for their poorer members. Their ability to organize protests and to close their shops gave them a certain amount of protection against government rapacity. A strike in 1123 forced a caliph not only to drop an unpopular tax but also to return the moneys already collected. The same tactic is used today: merchants can paralyze a town by closing their shops.

Other than through the guilds, little attention was paid to the "upward" push of politics. Orders came down from above with little attention to the desires or fears of the inhabitants. But, from time to time, we see attempts to ameliorate conditions. One was an echo, unintended no doubt, of Roman practice. What the Romans attempted in the office of *tribunus*, something like an ombudsman, the Baghdadis tried through an official called a *rais*.

The position of women in Baghdad was less open than among the earlier Arabs. Sequestering of women has a curious history. In roughly contemporary times, women in widely differing civilizations were segregated in either place or dress, or both. Japanese women were hidden behind screens; Byzantine women were usually veiled; and Western European women wore the clothing that until recently we associated with Catholic nuns. What seems to have happened in Islamic society was that the Arab invaders saw that aristocrats sheltered their women, whereas peasants allowed them to walk free. As they fell under Persian influence, the Arabs increasingly restricted their women. But, as late as the twelfth century, women in Baghdad gathered with men at least in mosques and probably in other public places. Those who wished to assert women's rights quoted a tradition of the Prophet Muhammad to the effect that in the law, men and women were equal; the tradition was patently false. However, quoting traditions was the customary way of justifying current behavior. So possibly women were freer than the existing chronicles tell us. Certainly women owned property and at least some played a role in politics. It is another of those characteristics of Iraqi history that Iraqi women became the most liberated in the Islamic East—well educated, engaged in all the professions, occupying senior government positions, and even serving in the armed forces during the twentieth century.

What we know of medieval Iraq comes from chronicles whose authors paid little attention to such "social" issues as the position of women but concentrated on the person of the ruler. The Abbasid caliph best known in the West, and indeed also among Arabs, is Harun ar-Rashid. The image we have of him comes mainly from numerous translations and movies derived from *The Arabian Nights*. His exploits, particularly his alleged penchant for wandering the streets of Baghdad in disguise to learn what the common people were doing and saying, became part of the folklore of later rulers. Even Saddam Husain was said to have tried occasionally to copy Harun.

What Harun did of more lasting consequence was to destroy the central command of his vast bureaucracy, the Barmakid family. Without them, the empire weakened. Province after province broke away under local dynasties that paid lip service instead of taxes to Baghdad. More important, Iraq's complex irrigation system was allowed to fall into disrepair: dams broke and canals silted. Without proper drainage, much of south Iraq became the vast swamp it remained until the end of the twentieth century.

When Harun died, the previously suppressed split between the Persian and Arab wings of Abbasid supporters came into the open in the struggle for power of his two sons. Even the son who was strong enough to win, Mamun, was not able to stop the slide toward ruin. In the twenty years of his rule, the revenues of the central treasury declined drastically. As they did, provincial landlords and tax farmers squeezed the peasantry ever more tightly, and to replace the former "citizen soldiers" who had created the dynasty, foreign mercenaries were imported. In less than a century, the real rulers of the empire were not members of the Abbasid dynasty but Turkish soldiers from Central Asia.

In 836, in an attempt to escape the very soldiers he had brought into Iraq and the hostility of the citizens who quickly learned to hate these foreigners, the Caliph Muatasim moved out of Baghdad and began to create a new capital, indeed a refuge, for himself and his court on the site of a settlement dating back thousands of years to Ubaidian times at Samarra and not far from the capital city created by Sargon. There he built a sort of Iraqi Versailles, with vast gardens, a mosque three times the size of St. Peter's basilica and palaces. Imperial grandeur was another of those themes that would be picked up by later rulers and carried to the extreme by Saddam Husain. Grand as Samarra was, it offered only a respite from unruly Baghdad. Soon caliphs fell like wheat to the sickle: in 861 the Turkish praetorians murdered a caliph and in short order imprisoned and then killed four of his successors. Samarra proved to be more of a jail than a refuge. In 892 the caliph and his court gave up and moved back to Baghdad.

While Baghdadis from caliphs to craftsmen were bossed by the Turkish soldiers, the peasants were left alone to dig and plow their small plots as they had done since Ubaidian times. Having little, they were not robbed. Remote and scattered, they were impervious to what happened in the capital. Their poverty would protect them for most of the next thousand years.

Very different was the experience of a new people who were brought into the far south. There, in the sweltering heat and high humidity, were minerals to be mined. To perform the heavy labor, Baghdad entrepreneurs began to import black slaves from Zanzibar. For the Zanj, as these people were known in Arabic history, life was miserable and short. They rose in revolt in the 870s. As with slaves in the Roman Empire a few centuries earlier, desperation made them competent soldiers. Not finally defeated until 883, they left a tradition of revolt against government and landlords that reverberated through the centuries down to our times.

In the decline of central administration and the loss of income from distant provinces, government revenues are believed to have fallen to only a few percent of those available to the first Abbasid caliphs. In a desperate effort to squeeze more out of the system, we hear of such macabre bits of bureaucracy as the "Office of Bribes" and the "Office of Confiscations." We have no contemporary records of how the common people reacted, but the record shows that they did all they could then and until the present to avoid any contact with government. For them, government always equated to taxes and often meant ruin.

From its position as one of the mightiest empires of the world, the Abbasid caliphate became itself an object of plunder.

Two groups deserve at least brief mention. From the northern Iranian highlands near the Caspian came the Buyid Shiis. Unlike the Turkish praetorians, they actually invaded Iraq and took Baghdad in 943. Under them the caliphate became a largely ceremonial institution resembling Momoyama Japan when its emperors similarly lived under the military elite we know as the shoguns. The Buyids were to dominate at least the capital of Iraq for more than a century until replaced by a new Turkish group, the Seljuks, who established an empire partially under the shadow of the caliphate.

Once again, as in the rise of Islam, Iraq was to be affected by events far afield. To understand why the Seljuks went to Iraq, we have to look right across Asia. Turks played a major role in Chinese history. It was a Turkish general who ended the Tang Dynasty in 907 and Turkish officers who established the "Five Dynasties" that preceded the Song Dynasty. The reestablishment of a strong, centralized rule in China in 960 by the Song closed the northeastern frontier to Turkish tribal incursions. Blocked in the east by the Chinese and under attack from the Mongol Ch'i-tan, the Turks turned west toward the Islamic world. Not as individual warriors, as in earlier centuries, they now came as whole tribes under their own leaders and intended to stay. In 1055 they captured Baghdad. For them Baghdad was only a camping ground; most of their activities were outside Iraq but in one area they left a lasting impact on Baghdad. It resulted from the actions of a Persian by the name of Nizam ul-Mulk, a prime minister to the Seljuk kings, who created one of the most impressive educational systems in the world of his time. He aimed to establish a college of higher learning in every significant city in the empire to train a competent civil service. His example stimulated others, and for a while there was a sort of mini-renaissance throughout Iraq and Persia. Nizam ul-Mulk was certainly one of the most extraordinary men of the Middle Ages, East or West. He planted in the minds of successive rulers an ideal of a government that could be measured by its dedication to education. Thus, even in the dictatorships of modern Iraq, leaders such as Abdul Karim Qasim and Saddam Husain applied themselves in his tradition.

Throughout these turbulent years, Iraq and particularly Baghdad evinced a remarkable tenacity. Rulers came and went; invasion followed invasion; floods drowned parts of the city and much of the countryside; dikes were broken and canals were allowed to silt. Mamun's "City of Peace" had little peace, but the city continued to thrive despite all that men did to destroy it. Recovering somewhat, the caliphate itself seemed to draw new energy from the city. But, unpredictably, infinitely worse was to come.

Far off across Central Asia, the greatest conqueror of all time was born in 1155. Genghis Khan almost did not survive a violent youth. He spent it fighting first for survival and then for control of nomadic Mongol tribes. In early adulthood, he launched his by then powerful armies across Central Asia and deep into China. In 1220, just seven years before his death, one of the Mongol armies swept into western Asia, capturing and destroying the great city of Bukhara. Next the Mongols took Samarqand, looted it, and enslaved what they regarded as the useful members of the population, its craftsmen. The rest they massacred. The story is told that one woman sought to buy her life by giving her captors a large pearl. They agreed and demanded the pearl. She replied that for safekeeping she had swallowed it. They immediately disemboweled her. Finding several pearls in her stomach, they reported the incident to Genghis Khan. He then ordered that all the bodies of the fallen be cut open and checked for hidden treasure. Everywhere they went, the Mongols cut down the people, burned or leveled towns, and destroyed irrigation works so thoroughly that for hundreds of years the areas they swept across remained deserted. Some have never recovered.

The horror of this invasion disheartened the people of Iraq: what could they expect but death and destruction? The same terror swept through Europe and, probably, through those portions of China not yet invaded. Except for the Russians, who were already conquered, the Europeans were lucky, and the Iraqis had a short reprieve. Genghis Khan died in 1227. Then the invasions began afresh. In 1251 Genghis Khan's grandson Mongke sent his brothers, Qubilai (Kublai Khan) to finish the conquest of China, and Hulagu westward into the Islamic countries. Hulagu focused on Baghdad, still one of the greatest cities in the world, with a population of about a million people. To beat down the walls, he brought a corps of Chinese siege specialists, but his main force was his disciplined cavalry armed with powerful compound bows. It was these mounted archers he turned on those troublesome peoples of the Iraqi-Persian mountains, the Kurds. Not for the last time would invaders seek their total destruction. Then he moved on toward Baghdad. There, Caliph Mustasim—whose name appropriately means in Arabic "one who seeks refuge"—asked for a truce. Hulagu's answer was to storm the city.

Three Mongol armies completely surrounded Baghdad in January 1258. Bravely but foolishly, the caliph's main force sallied forth from the walls. As they usually did, the Mongols pretended to retreat. Then they flooded the area behind the attacking Baghdadis, counterattacked, and slaughtered them. A few days later, the Mongols began their attack on the city, and in less than a week had breached its defenses. On February 10, in an incredible act of courage, Mustasim walked out of the city to surrender. He found no refuge. He was forced to order the inhabitants to lay down their arms and follow him. As they did so, the Mongols mercilessly cut them down.

Taking the caliph back into the city, Hulagu asked him where he kept his treasury. Then, probably something like the fanciful account of a chronicler actually happened. Upon being shown the contents of the treasury, Hulagu is said to have placed a tray heaped with gold before the terrified caliph and ordered him to eat. When the caliph replied that he could not eat gold, Hulagu taunted him for failing to use it to organize a better army. Resigned to his fate, the caliph is said to have mumbled that what he did was God's will. Well, then, said the Buddhist Hulagu, your fate is also God's will.

God's will as interpreted by Hulagu came swiftly. Mustasim was allegedly rolled in a carpet and stamped to death. This was the Mongol custom to avoid shedding royal blood. Mustasim's family and untold thousands—the contemporary guess was eight hundred thousand—of the townspeople were then killed, and the city was given over for a week of looting.

With that resilience we have witnessed in our own times in war-destroyed cities, those Baghdadis who were still alive set about recapturing as much as they could of their "normal" lives. What happened in Baghdad in the aftermath of Hulagu was comparable to what happened in Hiroshima after the atomic blast. Bricks were piled up again, goods were bought and sold, young people got married, children were born, life went on. But, unlike the people in modern disasters, the Baghdadis had no organized society to fall back on. For years following the actual massacres, starvation and disease took immense tolls.

The horror of the destruction and its aftermath still numbs the mind. Why did the Mongols do it? The best answer, I believe, is that they were trying to reduce the whole of Asia to the only kind of economy they knew, nomadic herding. For them, the cities, their populations, and, above all, their irrigation works were obstacles rather than assets. The vastness of China would eventually tame those who went south with Qubilai; Iraq was not large enough to tame those who rode with Hulagu.

After a turbulent century in which various rulers fought over the offal and the bones the Mongols had left behind, another worthy successor to Hulagu arrived on the scene. Baghdad was still high on his list of targets, although it had little of the riches or population of previous times. If Timur (Tamerlane) killed fewer, it was largely because there were fewer then alive. But he did the best he could with what he found. He is said to have slaughtered about ninety thousand Baghdadis in 1401. Like many of the great cities of western Asia, Baghdad had become a cemetery.

A century later, almost literally from the ashes, a new dynasty arose in Iran, the Safavids. As Shiis, they were attracted to Iraq because it was already the object of pilgrimage to the graves of their "saints." So the new Safavid shah made a triumphal tour of the holy cities and, to celebrate in the now accustomed manner, massacred many of the leading Sunnis.

It is useful here to dilate on the difference between what Sunni Islam and Shia Islam had become in Iraq and Iran. Mongol massacres were far more devastating than the catastrophe contemporary Europeans suffered in the Black Death. Destruction of the physical infrastructure was beyond repair. But, even more important, defeat after defeat at the hands of the Mongols and their successors occasioned a loss of that sense of divine favor that worldly success had imparted to Islam. This loss particularly affected the Sunnis, who for centuries had been supreme. Decline of Sunni prestige and power opened the way for the quite different thrust of mysticism. Whereas the Sunni rulers had emphasized law, formal theology, and rationalism, the common people in their misery and fear hungered for the comforts to be got from mysticism and otherworldliness. Sufism spread rapidly even among the Sunni upper classes. Pilgrimage to holy places, veneration for holy men, and the cult of saints virtually reshaped Islam. In short, what had always existed as folk religion came to the fore. It was, in part, this groundswell that explains the rise of the Safavids and their fascination with Iraq: the emotional thrust embodied in the passion of the *imam* Husain had been mingled with the mystical forces of the pre-Islamic Persian religion, Zoroastrianism.

Many of these tendencies also affected the population of the newly created Ottoman Empire. Sunni mysticism and Shiism spread widely throughout Anatolia. But they did not affect the Ottoman rulers. Traditionalists, they were completely out of sympathy with mysticism or otherworldliness; for them, God's message was embodied in order and authority. Thus it happened that when one of the most powerful and violent of the Ottoman sultans—known as Salim the Grim—came to the throne in 1512, he reacted to the rise of Persian Shiism and the Safavid ruler Shah Ismail's proclamation of it as Iran's state religion much as the Catholic king of Spain reacted to Queen Elizabeth's English Protestantism. The main difference was that Sultan Salim's "armada" moved by land.

Taking the shah's massacre of his Sunni coreligionists in Baghdad as an excuse and angry that Persians were inciting subversion in his empire, Salim decided to conquer Iraq and humiliate Persia. Against the religious conviction of the Persians—then as in the 1980s chanting a battle cry that proclaimed their desire for martyrdom (Arabic: *shahadah*)—the Ottomans employed cannon, disciplined infantry, and superior numbers.

In the battle of Chaldiran in midsummer of 1514, the Persian army, which had few if any firearms, was shattered. The Turkish victory was complete—the new shah narrowly missed capture when even his harem was overtaken by the Ottoman troops. But logistics were against the Turks. So both sides accustomed themselves to centuries of intermittent, wasteful, and unwinnable frontier wars in which Iraq was the spoil and the battleground.

This long period of Turkish-Iranian hostility washed over the peasants and tribesmen but affected urban Iraqis in two ways: it slowed or prevented economic recovery, and it solidified the division of the population between Sunnis and Shiis. Urban Sunnis tended to identify with the Ottoman government, while urban Shiis identified with the Iranians. But the differences were far more profound than this statement suggests. Since they were generally the suppressed party, the Shiis drew tightly together as a community; in it, religion and politics were united under "guides," as the Persian Zoroastrian tradition mandated. These jurisconsults of religious law (Arabic: *mujtahids*, literally those who go deeply into matters) formed a collective leadership known as the authority (Arabic: *al-marjiyah*) that took upon itself the tasks of embodying the tradition of the community, teaching their successors and guiding their flock. While separate from government, they actually carried on many of the functions we ascribe to government, running the educational system, collecting taxes, judging disputes, and issuing legal decisions (Arabic: *fatwas*). Since they looked upon Sunni government as illegitimate, they sometimes employed their own quasi-military forces to defend themselves or to attack others. They still do. In the eyes of the government, of course, such actions constituted terrorism and were often ruthlessly suppressed. This conflict continues in present-day Iraq, where the *marjiyah* is a shadow government.

Nothing comparable to the *marjiyah* existed in Sunni Iraq. It wasn't needed. The Ottoman government spoke in the name of all Sunnis. But it spoke with a muted voice. Having reverted to the primitive, almost pastoral society of Ubaidian times, Iraq was too poor to be of much interest to the empire. So the Turks thought of themselves as the shepherds and the Iraqis as their flock. Without wasting resources that they needed for their invasion of Europe and their control the far richer land of Egypt, they sought merely to shear the sheep. ("Sheep" was actually the term they employed.) Otherwise, they spent as little money and effort as would maintain a minimum sense of order.

Toward the middle of the nineteenth century, Iraq began to undergo a number of changes that would permeate its society for the next century. The most important of these was the transformation of "tribal" society from seminomadism to settled agriculture. The Ottoman government promoted this change because farmers were easier to control (and tax) than bedouin: the "sheep" were easier to shear when confined to a smaller area.

To facilitate investment and tax collection, a reforming governor, Midhat Pasha, attempted to codify the various rights to land that had accumulated through the long sequence of kingdoms and empires and that were superimposed haphazardly on customary rights that went back into the first Ubaidian settlements. Midhat's work was given impetus by the growth of the export trade, particularly in rice and dates, which encouraged city merchants and tribal leaders to invest in land reclamation, canal digging, and dam building. Those with marginally better skills and/or connections quickly found ways to increase their holdings and, with the occasional help of government soldiers, their control over the formerly independent tribesmen. This began a process that the British encouraged during the First World War and that was to reach its culmination during the 1930s, when the tribesmen became virtual serfs.

As cash crops came into vogue, agricultural land, particularly in the well-watered south of Iraq, doubled in extent. Then, lured by a new sense of regularity of life—bluntly put, by the possibility of having full bellies throughout the year—large numbers of bedouin, most of whom had already raised some crops when sporadic rainfall had made that possible, gave up their nomadic life to become full-time farmers. Fearing government and city merchants, both of whom were Sunni, many became Shiis. They are the ancestors of the southern Iraqis of today.

Foreigners had little to do with these developments. As they began to arrive or pass through Iraq in the eighteenth century, they found Iraq too poor to be of interest. The British, however, discovered a use for it: the Euphrates formed a link in the route from their growing empire in India to England. For this link, they recreated the system employed by the ancient Persians and Mongols, a postal service mounted on camels. Using the "British Dromedary Post," Englishmen in India could communicate with London in a matter of weeks instead of the months required to sail around the Horn of Africa. Following the Euphrates was also safer than risking the shoals of the Red Sea in ships powered only by sails. To maintain the route, the British set up consulates in Basra in 1764 and Baghdad in 1798; they kept the route reasonably safe, but the post still was often delayed and sometimes failed to arrive. In 1800 a senior British official in India complained that he was "nearly *seven* [his emphasis] months without receiving one line of authentic intelligence from England . . . Speedy, authentic, and regular intelligence from Europe is essential to the conduct of the trade and government of this empire." So a few years later, in 1834, the British imported the first steamboats on the Euphrates and Tigris.

What the British began, other Europeans took up. From about 1840, a scheme to build a railroad began to be discussed. A railroad would have solved the problem of speed, but it made no economic sense, given the small Iraqi population. So, in 1872, an Austrian engineer suggested a solution: settle 2 million Germans along the Euphrates. That was not done, but, to the horror of the British, in 1899 a concession to build the railroad from Istanbul to Basra was given to a German company.

Fear of Europeans advancing toward India along the Euphrates had been a British nightmare since 1798 when Napoleon landed thirty-eight thousand French soldiers in Egypt. A year later, he began the conquest of Palestine and Syria as stepping stones toward India. He was defeated by an outbreak of plague. Realizing that they could not always count on plague, the British sent an army to hasten the French withdrawal. But their real worry was not that the French would keep Egypt so much as that they would move into Iraq. As early as 1798, the British secretary of war wrote that "Bonaparte will, as much as possible avoid the dangers of the Sea, which is not his element, but . . .by marching to Aleppo, cross the Euphrates, and following the example of Alexander, by following the River Euphrates and the Tigris, and descending to the Persian Gulf will march on India."

As the French challenge receded, it was seen to be replaced first by a Russian threat (which was mainly played out in the covert war known as the "Great Game" in Afghanistan but also affected Iraq through events in the Ottoman Empire) and then by Germany. When Germany was united in 1870, it undertook an aggressive foreign policy. In the Iraqi part of this policy, it pressed for a railway concession in 1899—the "Berlin to

Baghdad" line—and began a steamship service to the Gulf in 1906. Alarmed, the British forbade the shaikh of the little trading port of Kuwait from giving the Germans any facilities for their ships and began the process that would make Kuwait a separate state. Kuwait was to be the "cork" in the Iraqi bottle. To be sure that the cork stayed tight was the essence of British policy in the Gulf.

But events in Iraq soon created two new aims for British policy. Discovery of a huge deposit of oil in Iran, to the east of Abadan, in 1907, led the British to believe that Iraq might also contain oil. Every reader of the Bible knew of the "fiery furnace" that, surely, was in Iraq. And with the Royal Navy just converted to burning oil instead of coal, the British had conceived a "vital national interest" in oil. Oil was the wave of the future: those who controlled it, the British believed, would rule the world or at least not lose the empire. As the eminent British statesman Lord Curzon later said of the First World War, "the Allies floated to victory on a wave of oil."

Britain also began to think of another possible value of Iraq. It is one that again takes us back to the beginning of our story with the agricultural revolution: "Babylonia," they thought, might indeed again become the Garden of Eden. On the eve of the First World War, one of those eccentric Englishmen who were drawn to the East, Canon J. T. Parfit, called it "the key to the future." The most influential adviser to the British government on the Middle East at the time, Sir Mark Sykes, wrote that "There is no doubt that the land [of Iraq] is the richest in the world." Even sober engineers became nearly poetic when they described Iraq. The leading English authority on irrigation wrote in 1910 that as soon as the Tigris and Euphrates were brought under control, the millions of "surplus" Indians then dying of famine in British India could be settled along the Tigris and Euphrates. There, as farmers, they could produce as much wheat for food and cotton for industry as required by the whole British Empire. Iraq could, he said, "attain a fertility of which history has no record." And so Iraq was poised in the summer of 1914 to enter a new era, to become "British Iraq."

THREE

BRITISH IRAQ

Although the First World War had begun in Europe in August 1914, Britain did not declare war on the Ottoman Empire until November 5. Before the formal declaration, however, it recognized Kuwait, then a part of the Ottoman Empire, as an independent state under British protection. Then on November 6 it landed a British-Indian military task force at the southern port of Fao. Advancing inland, this force secured the area around Basra. The British took these steps ostensibly to protect the oil field in nearby Iran whose production they needed for their navy, but from the first days, their occupation was marked by a very different objective. They immediately began to introduce British-Indian laws, police, bureaucracy, and government in the area of their control. That is, they began to treat their piece of Iraq as a part of their Indian empire.

Why did the British decide to conquer Iraq? Contemporary answers to that question are as obscure, convoluted, and tendentious as justifications offered for the American invasion of Iraq in 2003. Since they shaped much of the future, they deserve to be clarified.

In the diplomatic papers that passed between London and

Delhi in the years before the war, the threat of what was then called "Pan-Islamism" figure prominently. The Allies—Britain, France, and Russia—dominated huge Muslim populations in Africa and Asia. Each feared that its subject Muslims might try to drive them out. The Sepoy Revolt of 1857 provided the text that British statesmen of 1914 had studied. They dreaded what might happen in India in the event of war with the Ottoman Empire, in the name of whose sultan-caliph prayers were offered daily by millions of Indian Muslims. Would they rise against his enemy at his call? Their British rulers feared that they would. They always believed that the masses of Asia were on the point of rising. Their security forces told them, as policemen are prone to do, that plots were being hatched almost daily all over India. Since the British officials already *knew*, "intelligence" was irrelevant.

If they were in any doubt, the "worst-case scenario" was backed up by what the British heard from the Russians. The tsar had several times discussed with the British ambassador his concern about Turkish "Pan-Islamicist" propaganda among the "seething" Muslim rebels in Russia's Central Asian empire. All was rumor, and much turned out to be myth, but apprehension was great. That apprehension may perhaps be gauged best not in sober diplomatic dispatches but in a then-popular novel by John Buchan. *Greenmantle*—precursor of the James Bond series that later was to captivate President John F. Kennedy—excitingly cast sinister Turkish and German agents promoting a holy war that was thwarted only by intrepid British agents. *Greenmantle* gave us "007" long before Ian Fleming invented him.

Leaving aside bizarre but enjoyable fiction, there was, of course, some truth in British and Russian fears. There always is.

Myth becomes significant only when laced with a dash of fact. But the real job of analysis was then, and remains today, to decide how much fiction gets mixed with fact. Once the responsible British statesmen made some sense of what they were hearing, their task was to determine what they could do about it, what the alternatives might be, how likely each was to succeed, and what costs each would involve.

By the time they had begun to get a grip on the basic facts, their alternatives had narrowed. The Germans had won over the Turks. Before the war actually broke out, the British might have worked more effectively to keep the Ottoman Empire neutral. That would have been a difficult task, of course, since they and the Russians had been attacking the Ottoman Empire for a century. But the cost of *not* keeping the Ottoman Turks out of the war was not properly evaluated. It turned out to be nearly catastrophic.

When Britain precipitated the war with the Ottoman Empire by its invasion of Iraq, a number of actions were bound to follow. The Ottoman sultan responded a week after the British landing in Iraq with just what the British, French, and Russians most feared, a call for a holy war (*jihad*). He was bound to do so. The British and French were lucky, but it was dumb luck. There was no massive uprising among their subject peoples. Luck did not save the Russians, however. Russia could neither feed nor arm its huge armies. So when the Turks closed the supply route through the Dardanelles and the Bosphorus, Russia began to starve and its armies began to collapse. The Revolution of 1917 then became almost inevitable. That event released a whole German army to fight on the already hard-pressed western front.

Other policy failures followed. Having decided to go to war, the British could have accomplished what was really important to them, protecting the oil production of Iran, by occupying only the small area around Kuwait and Basra. Instead, they decided to take what they called "that tract of country known as Mesopotamia." In June 1915 the little army they had put ashore at Basra imprudently advanced toward Baghdad. Why did they do this?

A reason never to be discounted in wars is that generals are hired to fight. That is how they win promotion and decorations. Sitting in Basra was tedious. They obviously felt overshadowed, bypassed, marginalized by the great events unfolding on the western front. They did not want to miss their moment in history. Moreover, whatever they wanted, they were not totally in control. Once begun, military operations tend to develop a momentum of their own and are hard to stop. As the chief political officer of the British force commented, once the troops had landed at Basra, he did not see how "we can well avoid taking over Baghdad."

It wasn't just the momentum of the action or the ambitions of the generals; there were then, as there always are, other justifications. Intelligence reports indicated that the Turks were concentrating troops and preparing to attack. In the background, special interests were pressing for their advantages. The generals were being pushed by British merchants, who saw an advantage to them in the advance. Americans would feel the same pressures in Iraq nearly a century later.

In any event, the British believed defeating the Turks would be easy. The Ottoman Empire would collapse at the first touch. European statesmen had long scoffed at this "Sick Man of Europe." Again, they misjudged. The Turks were brave and hardy soldiers. The war would prove, as wars usually do, far more costly and difficult than anticipated. At Gallipoli, the

Turks fought half a million British and French troops to a bloody standstill. The failure of that campaign was what starved Russia into collapse.

Of less strategic importance but also costly was the result in Iraq. When the British marched toward Baghdad, the Turks drove them back and trapped a whole British division in the little town of Kut a hundred miles south of Baghdad. For four months, the British tried to break the siege. In that effort, they lost 7,000 soldiers. In desperation, they tried to bribe the Turkish commander to let the trapped troops go free. Insulted, he refused, and his troops forced the surrender of 13,309 British soldiers. Conquering Iraq would take almost four years and cost another 20,000 British (mainly Indian) casualties.

Viewed in terms of overall strategy, more significant than casualties was the diversion of troops. Protecting the Suez Canal, which the Turks attacked in the spring of 1915, and driving the Turks away from the southwestern Iranian oil pipeline, which they cut for three months in 1915, required more than a million British soldiers who were desperately needed on the western front. The monetary cost also was staggering. The Middle East drained away the then vast sum of £750 million.[*] None of this was anticipated or even considered when the decision was taken to invade Iraq. Again, we find parallels to events of our own times, when the amount of money we were told the Iraq war would cost, $30 to 40 billion, is only about 10 percent of the probable net amount.

Perhaps the British would have invaded anyway. Put against

---

[*] In terms of currency, this is roughly the equivalent in today's terms to $18 billion, but it was a much larger percentage of Britain's then gross national product than the number suggests.

the costs, there were real assets to be gained in Iraq. It was thought almost certainly to contain oil and its agricultural potential was believed to be enormous. Both oil and food would increase the power of the British Empire and make its Indian possessions more secure. But Britain was under great financial pressure so the trick was to rule Iraq on the cheap. To save money and shipping space, the army should live off the land. Every effort began to be made to stimulate local production of foodstuffs. This effort was partially successful. By the end of the war, the British managed to bring into production 500 square miles, or 1,300 square kilometers, of formerly desert lands on which 50,000 tons of grain were produced. They also promoted the manufacture of simple equipment and the salvaging and repair of what the army had worn out. More important, they decided to rule Iraq with a "lean" military force. Actually controlling the population should be accomplished by a specialist corps of political officers with some help from locally recruited guards. This, after all, was what they had learned to do in India. Since the retreating Turks took away their records and evacuated most of their officials, they imported Indian clerks to keep the books.

As the costs of the Iraq campaign began to mount, and were added to the huge expenditures of the war in Europe, it became evident to the government in London that additional ways to economize must be found. During the early years of the war, however, this message did not get through to the British officials in Iraq. As long as they were actually fighting the Turks, they could justify what they were doing. But their war ended in armistice on October 30, 1918.

The best way to save money, the London government decided, was to reduce the need for an army of occupation. So, despite his protest, the commanding general was ordered to read a proclamation to the citizens of Baghdad inviting them "to participate in the management of your civil affairs in collaboration with the Political Representatives of Great Britain." This statement baffled those Iraqis who heard it. Did it mean that they were about to be independent? Did it mean that they were becoming a colony? Did it really mean anything? No Iraqis could tell. Nor could their British masters. The proclamation simply papered over profound differences among the British officials and disguised their lack of decision on policy.

At one end of the spectrum of possible British policy was one of the most extraordinary Englishmen of that era. Arnold Wilson was a brave soldier, winner of one of Britain's highest decorations, a scholar on the languages and cultures of Iraq and Iran, author of memoirs that are a model of its genre. Of unchallenged personal integrity, he ran an efficient, effective, and honest administration. His aim was order and economy. But he did not expect advice and would not tolerate opposition from the natives. Their proper role was to acquiesce while skilled British officers, ruling like Plato's philosopher-kings, did what was best for them.

As Wilson and his staff—known among the English as "the Indian School" since all had come to Iraq from service in Britain's Indian empire—saw it, the Iraqis divided into three groups. The first group was made up of the bedouin and Kurds. They were rather like the Pathans of the Indian Northwest Frontier province, fierce and colorful, admirable but dangerous, noble savages who could not possibly run a modern state. The second group was similar to the vast Indian peasantry, pitiful, poor, ignorant, and obviously incapable of governing themselves. The third group was the worst. They had just enough capability to be truly destructive. They were the "town Arabs." If let into government, they would ruin the whole country. The bottom line was that the British must rule Iraq. Any other view was simply naïve and irresponsible.

Wilson was sure that sensible Iraqis agreed with him. On November 16, 1918, he wrote a dispatch to London and Delhi that with a few changes of address and substitutions of names might have been written in 2003 to Washington: "The average [Iraqi] Arab," he said, "as opposed to the handful of amateur politicians of Baghdad sees the future as one of fair dealing and material and moral progress under the aegis of Britain. . . . The Arabs are content with our occupation."

Wilson would soon find how wrong he was. When Iraqis began to hear that Britain had decided to award itself a "mandate"—which they took to mean colonial status for them—they began to gather in the only public places they knew, the mosques, to hear sermons against the British. When a "committee of delegates" (Arabic: *mandubin*) tried to present a petition to Wilson, he first refused to see them, then packed the meeting with carefully selected supporters and, to top it off, stationed a gunboat in the Tigris with guns trained on the meeting place. Then he heard what he wanted to hear from what they said.

It was to be years before the British admitted what had been clear even then to neutral observers. The Iraqis did not want Britain to run their country. Those Englishmen who understood thought that they had to wear a veil to hide the face of their rule. The veil was to be supplied by the League of Nations. Iraq would not be a British colony but would be "mandated" to Britain by the League so that Britain could prepare its people for self-rule. The plan was drawn up by a group of English officials including Colonel T. E. Lawrence—known among the English as "the Sharifians" since they had been working with the family of the sharif of Mecca in the "Revolt in the Desert"—in the spring of 1919. The mandate was ratified by the principal allied powers at a conference at San Remo in April 1920. To many Iraqis, used to guessing what was behind the veil, it seemed proof that Britain meant to stay in Iraq.

They were profoundly disturbed. As the official British report of 1928 finally acknowledged, "From the beginning, the idea of a mandate has been abhorrent to nearly all educated elements in the country." While most opposition Britain encountered, as this report made clear, was native to Iraq, there was some outside agitation. That fact made it possible for the British officials then, as for American officials later, to allege that domestic opposition was weak, composed of disaffected "remnants" who were being incited by foreigners. Let us examine that because it was important then and later.

Since no agreement had been reached on the frontiers, the position of Mosul was unclear: was it to be Iraqi, Syrian, or even Turkish? All along the Euphrates, particularly in the little town of al-Qaim, where fighting would erupt again in 2003 and 2004, the claims of "nationalists" were pushed aside by the British. Then, when the British were ordered by London to calm the situation by pulling back from violent confrontation, the nationalists thought they had achieved a victory. Finally a small group of nationalists, acting in the name of the wartime secret society, the Iraqi Compact (Arabic: *al-Ahd al-Iraqiyah*), and pompously calling themselves the "Northern Iraq Army," tried to capture Mosul in May 1920. In a foretaste of events in 2003, they managed to ambush two armored cars and shoot down a military aircraft.

In the primitive communications of the time, probably few Iraqis heard of these events, and the members of the "Northern Iraq Army," few as they were, were soon arrested or forced to flee. Their pathetic activities were completely stopped when the

French invaded and conquered Syria. The British could relax. They had won. A battle, that is, but not a war.

What counted to most of the Iraqi population were not outside agitators but events inside Iraq. These have not been well studied and deserve attention both because they deeply affected the Iraq of that time and because they have a current relevance.

As they had done on the Indian Northwest Frontier, British political officers were trying to achieve "security" by what appeared to them a logical intervention into society. Believing that Iraqi tribes were like Pathan tribes, they identified local notables who would work with them and "promoted" them to be "chiefs." They then codified the position of these men in "Tribal Disputes Regulations," which gave the new chiefs an authority over their kinsmen that was revolutionary. To underwrite this new authority, they gave the chiefs various advantages, including, as they specified, "large doles, subsidies and no taxation," along with confirmation of their private ownership of lands that had previously been regarded as communal to tribes. Given these new powers, the British assumed that the chiefs would have a stake in British rule and that they could control their fellow tribesmen. But what they stimulated was a social revolution within what was growing into a nationalist revolt. The new chiefs excited the anger of their fellows when they used their positions both to acquire riches and to serve as the advance force of a British occupation that would become permanent.

As anger intensified, the British reacted with armed force. This, in turn, had the effect of driving the formerly mutually hostile Sunnis and Shiis closer together—again as would happen in 2003 and 2004. During the Holy Month (Ramadan), joint Sunni-Shii meetings were held at which appeals were sounded for unity against the British. As was traditional in times of danger, the bazaars closed; British soldiers were attacked; and the leaders of the Shia community issued instructions to their followers, particularly among the tribes of the south, to rise against the British. On June 30, 1920, Iraq blew up in a vast insurrection against the British.

The British then had 133,000 troops in Iraq. Arrayed against them at any given time was only a fraction that many Iraqis. With uncanny accuracy, the number of both troops and partisans are almost exactly those of 2004. But unlike the Americans in 2004, the British troops in 1920 were widely scattered and relatively immobile. For guerrilla warfare, regular troops were not suitable, whereas hit-and-run tactics were the metier of the tribesmen. For the next six months, as the British battled against virtually the whole population—even including the supposedly anti-Arab Kurds—they lost 1,654 men and spent more than six times as much as they had spent on the whole of their wartime campaign in the Middle East.

The British government was horrified. This was no tribal revolt. It was a national war of independence. Tribesmen did much of the fighting, no doubt, but they were led by respected men of religion, both Sunnis and Shiis, doctors, teachers, merchants, journalists, and even those "tame" Iraqis who were being trained to be government officials.

The man who, above all others, Wilson thought to be naïve and wrongheaded, a flamboyant newcomer who had captured the ear of senior officials in London, fired the most telling shot in the debate. Colonel T. E. Lawrence ("Lawrence of Arabia") wrote a sarcastic letter to the *London Sunday Times* in August 1920, when Britain was trying to consolidate its own conquest of Iraq: "The people of England have been led in Mesopotamia into a trap from which it will be hard to escape with dignity and honour. They have been tricked into it by a steady withholding of information. The Baghdad communiqués are belated, insincere, incomplete.... We are today not far from a disaster...." Then, comparing the British experience with that of the then despised Ottoman rule of Iraq, he continued, "Our government is worse than the old Turkish system. They kept fourteen thousand local conscripts embodied and killed a yearly average of two hundred Arabs in maintaining peace. We keep ninety thousand men, with aeroplanes, armoured cars, gunboats and armoured trains. We killed about ten thousand Arabs in this rising this summer. We cannot hope to maintain such an average: it is a poor country, sparsely peopled."

Lawrence also put his finger on the sorest point of all: the monetary cost. Not only did Wilson's policy not work, it was too expensive for the war-weary English public to tolerate. Wilson had to go. In October he was replaced, and the British decided to implement their plan for the mandate. That plan called for Iraq to become "an Independent State under guarantee of the League of Nations and subject to the Mandate of Great Britain...." Just when this would take place was left vague. That is, British rule would continue "until such time as it can stand by itself, when the Mandate will come to an end."

The conditions for such a quasi-colony, quasi-independent state seemed hardly propitious. British troops were still locked in combat throughout the country, and the insurgents were still carrying out spectacular raids. Government archives had disappeared, and destroyed government buildings were still smoldering. As a sort of fig leaf to cover the nakedness of whatever Iraq was to be, the new civil commissioner, Sir Percy Cox, decided to set up a handpicked, provisional "Council of State." In a move that again was to presage American actions eighty-four years later, he appointed the Iraqi members. They were to operate in restricted areas, with limited powers, and to accept the "advice" of British officials. As titular and temporary "head of state," Cox appointed an aged Iraqi, the traditional figure known as the "Leader of the Descendents of the Prophet" (Arabic: *Naqib al-Ashraf*), under whose name the rebels could be amnestied.

In a then little considered aspect of his moves, Cox decided that the people he could work with in Iraq were the Sunnis. He rejected the offer of the leading Shia "clergy" (Arabic: *mujtahids*) of Najaf and Kerbala to negotiate an end to tribal unrest. Cumulatively his early moves would alienate the Shia community both from the British government and, subsequently, from the Iraqi government. The trend it set in motion has had profound implications down to our times.

Then, as the country calmed, exhausted by the rebellion, the British began to organize departments to administer rudimentary services.

What they had done so far was a series of stop-gap measures. Some permanent arrangement seemed to be required not only for Iraq but for the entire British area. So, in March 1921, the then colonial secretary, Winston Churchill, met with all the senior British Middle Eastern specialists in Cairo to lay out the organization not only of Iraq but of the entire British Middle East.

The principal question on the Iraqi part of the agenda was how to reduce expenditures. The military must be drastically cut. That could happen only if influential Iraqis could be persuaded that they had a government. Churchill agreed. But who would lead it? Two local candidates had already been evaluated and rejected. The Naqib al-Ashraf, head of the interim government, was dismissed as too old. The other was the interim British-appointed minister of interior, Sayyid Talib, a man who had rendered great services to the British during the war and the later rebellion. A man of undoubted ability,* he was the most popular notable in Iraq and was gathering supporters throughout the country. What made choosing him impossible was that he had coined the slogan "Iraq for the Iraqis" (Arabic: *al-Iraq li'l-Iraqiyin*). When he spoke openly in front of an English journalist of his hope that Iraqis might choose their own leader, the British arrested him, bundled him into an armored car, and spirited him away to exile.

A few other candidates were discussed, but by the time of the Cairo meeting, it was already clear that a decision had been made to offer the crown to the man the French had recently exiled as king of Syria, Faisal, with whom the British had worked during the Arab Revolt and to whom they began to pay a subsidy.

Churchill knew little and understood less about Faisal. As he got ready to report to Parliament on the results of his grand settlement of the Middle East, Churchill made an astonishing request to his Colonial Office aide (on June 14, 1921) that showed the profound ignorance of Iraqi affairs that has marked high-level discussion of them ever since. "Let me have a note in about

---

* In 1910 the British intelligence officer reported that "a debauched but energetic Vali [governor], Suleiman Nazif, was sent to Basra. Sayid Talib wanted to get rid of him." So he "hit on the brilliant idea of persuading all the pimps, bath boys, dancing girls and keepers of houses of ill repute in Basra with a few noted bullies and cut-purses to send a petition to Talaat Bey, then Minister of Interior, praying him to maintain 'this energetic, affable and generous Governor' in his post and thus confute the machinations of the evil disposed and gladden petitioners' hearts. Talaat showed the petition to certain holy men (Arabic: *ulema*) of Basra and asked them if they knew the signatories. They replied that they were only too well known and Suleiman lost his place. This is an interesting example of his methods."

three lines," Churchill asked, "as to [King] Feisal's religious character. Is he a Sunni with Shaih [*sic* for *Shia*] sympathies, or a Shaih [Shii] with Sunni sympathies, or how does he square it? What is [his father] Hussein? Which is the aristocratic high church and which is the low church? What are the religious people at Kerbela? I always get mixed up between these two."

Faisal, who was widely known in Syria, Palestine, and Egypt as the leader of the Arab Revolt, was little known in Iraq because Iraq had been largely cut off from the rest of the Arab world during the war. Neither the British then nor the Americans 83 years later could find a person in Iraq they felt was suitable. So just as the Americans in 2003 focused first on Ahmad Chalabi and then on Iyad al-Allawi, neither of whom had been in Iraq for decades, the British imported Faisal. In the attempt to create a following for him, the British mounted a campaign to create a demonstration of favorable public opinion. Despite British efforts, however, Faisal was disappointed by the "lukewarm" reception he received.

Even more fundamental, though then perhaps less urgent than the choice of a ruler, was the actual makeup of the new state. What was "Iraq"? Under the Ottoman Empire, the area known as Iraq or Mesopotamia actually was divided into three provinces (Turkish: *vilayet*), focused, in the Ottoman custom, on towns: Basra in the south, Baghdad in the center, and Mosul in the north. Before the Russian Revolution, Britain planned to turn Mosul over to France, thus creating a buffer zone between their Iraq and a Russian-dominated Turkish-Kurdish-Armenian zone in the north. That way, the French would have to cope with the Russians. But when revolutionary Russia dropped out of what was called the Sykes-Picot plan to divide up the Middle East, the British changed their minds. Mosul began to seem worth keeping when they realized that it was likely that the area contained vast deposits of oil.* So the British joined Mosul with Baghdad and Basra to form Iraq.

What to do about Kurdistan was a more complex issue. During the First World War, Russian troops had briefly reached Rawanduz in what became Iraqi Kurdistan; later, they tried to organize a base for themselves in the short-lived republic of Mahabad. These actions raised the old specter of an attack on the British Empire in India, and the British sent an expeditionary force to try to seal off the roads south from the new Soviet state. At the Paris Peace Conference, the British reluctantly agreed to set up a commission to devise a scheme for an autonomous Kurdish area. If the Kurds could demonstrate that they were ready for independence, it was to be given to them by the Council of the League of Nations. If that happened, then the Great Powers were to acquiesce in allowing the Kurds in the old Mosul *vilayet* to join the new state.

Sir Percy Cox and his supporters among the "Indian school" in Iraq insisted that Kurdistan be part of Iraq. Churchill wavered. He felt (rightly as it turned out) that an Arab Iraq would oppress its Kurdish minority. A compromise of sorts was reached: the Kurds would be temporarily kept apart but would be eventually integrated. What would ultimately determine the fate of Kurdistan had little to do with the Kurds; it would be decided by the fact that a huge deposit of oil was known to exist in what might have become a separate Kurdish state rather than British-controlled Iraq. The Treaty of Lausanne, signed July 24,

---

* The French were outraged. The British bought them off with a quarter interest in what came to be known as the Iraq Petroleum Company (IPC) and agreed to accept them as the mandatory power for Syria.

1923, recognized the Turkish state but made no mention of the Kurds. Oil made Kurdistan Iraqi.

Already at the Cairo Conference, the question of pacification of Iraq had been discussed. The RAF commander, Air Marshal Trenchard, proposed a revolutionary means to cut costs. Aircraft, he wrote, had proven themselves during the war. Machine guns, bombs, and poison gas (which Winston Churchill and the chief of the Imperial General Staff had urged be considered "a [legitimate] weapon of war" in 1920) would awe the nomadic tribesmen. Furthermore, aircraft could scout vast areas, detect unusual concentrations of people, and report to ground units. Then trucks mounting machine guns, called "armed Fords," could rush to the spot. Bedouin armed with aged rifles and mounted on camels could not stand against them. If they tried to do so, they could be attacked, as sometimes they were, with poison gas.

Control of the desert, the problem the Sasanian Persians had met by subsidizing the Lakhmid kingdom of Hira, was to be dealt with by aircraft. The Report of the High Commissioner to the League of Nations in 1923 summed up the effect: "a main factor in the pacification of the country," it said,

> has been the Royal Air Force. By prompt demonstrations on the first sign of trouble carried out over any area affected, however distant, tribal insubordination has been calmed before it could grow dangerous. . . . In earlier times punitive columns would have to struggle toward their objectives across deserts or through difficult defiles, compelled by the necessities of their preparations and marching to give time for their opponents to gain strength. But now, almost before the would-be rebel has formulated his

plans, the droning of the aeroplanes is heard overhead, and in the majority of cases, their mere appearance is enough. By its means (air power), it has been possible to achieve a highly centralized yet widely understanding intelligence which is the essence of wise and economical control."

Thus, already, the British had set in motion a revolution in the relationship of the urban and tribal elements of Iraq. Had this not been the case, the various political movements of the 1920s and 1930s would now be looked on as merely a continuation of the dreary series of court intrigues in Baghdad of the Ottoman era. The military aspects of establishing this revolution were to be seen not only in the work of the RAF, but also in the use of people whom the Iraqis regarded as "non-national."

There were then about twenty thousand "Assyrian" Christian refugees from Anatolia camped near Mosul. Enrolling the men in a military force, the British decided, would have the double advantage of feeding them and creating an inexpensive military force to replace some of the departing British military units. That force would have to be loyal to the British because the Iraqis would regard them as British surrogates. Backed up, "stiffened" in the modern military term, by "armed Fords" and planes from the RAF, the "levies," as they were known, would work for the British for a generation. By 1925 they numbered seventy-five hundred men.

Meanwhile, a new Iraqi, almost entirely Sunni Muslim, army, commanded at first by former Turkish-trained Arab officers, was also established. It was kept numerically equivalent to the levies. With the coming of formal, or ostensible, independence in the 1930s, this minuscule but uniquely armed, reasonably organized, and relatively mobile force would destroy the Assyrian levies in the summer of 1933. It would intervene time after time in political affairs.

As forces for keeping order in the 1920s, the levies and the army were of limited use. More significant but more subtle was the effect of the opening of "farm-to-market" roads. A British political officer in Kurdistan remarked in 1928 that

> a tribal chief from the vicinity of the Persian border having business at administrative headquarters would make the two days' journey accompanied by a large escort of armed horsemen. Following the construction of a pioneer motor road, with police posts, from Sulaimaniya, a regular taxi service has sprung up. The tribal chief, finding that he can take a seat for three rupees and perform the journey without fatigue in two hours, ceases to entertain large bodies of expensive armed retainers. The practice of carrying arms thus tends to grow less.

The roads also promoted a range of other changes. Finding that they could sell certain crops in the towns and buy what they could not themselves produce moved villagers toward dependency on outside markets in place of the autarky that had existed for thousands of years. As they had watched British soldiers use kerosene lamps and pocket knives, witnessed the effects of more modern rifles and shotguns, and were urged by their womenfolk to acquire bright cotton cloth, they rushed to market. There they indulged in chinaware from Japan, cigarettes from Baghdad, and shotgun shells from England. From the Iraqi copy of a British soldier's "overseas cap" downward, Iraqis began to put aside traditional clothing in favor of Western dress. Old things, old tools, old weapons, old habits began to be discarded. One by one, people came to have a stake in commerce, and through commerce a stake in public order, and through public order a stake in the state.

To pay for coveted goods, a new agricultural revolution was

required. A farmer could no longer wait for the annual floods. They were never very satisfactory since they came at the wrong time for the growth of plants. This had been a problem with which farmers had struggled since the first Ubaidian settlers began to work the land. Now an alternative was available. Water could be pumped onto the parched land. That alternative, however, could not be implemented until those who had money, the city merchants, bought the pumps; that they would not do unless the title to land was secure. So the British began to regularize ownership. The basis on which they built was the Ottoman code of 1858. That code favored those who would pay taxes: city moneylenders and notables.

Those who were disfavored were the "little people" who actually plowed the land, dug the canals, and cleared the ditches. Their claim on the plots they worked was customary rather than formal. For thousands of years their ancestors, just like European peasants, had feared government and sought to avoid it whenever possible. To a man they were illiterate. They did not understand or even know about whatever documents might exist in far-off cities. As long as their labor was the only thing that counted, they were safe. But with the advent of mechanical pumps and tractor-driven plows, they were in danger. That danger came swiftly.

In 1925 the British high commissioner reported to the League of Nations that "all lands excluding urban freehold (Arabic: *mulk*) properties belong primarily to the State and that good title to such lands can only be obtained in consequence of alienation by Government. . . ." What this meant was that the peasants had no rights to lands on which their ancestors had lived from time immemorial and "alienation by Government" effectively meant that those who had won government favor were now their landlords.

During the period from 1920 to 1932 great tracts of land along the rivers thus became private property. Economically, the result was at least initially positive. Production increased as the area of pump-irrigated land rose from just 72 to about 4,000 square miles, or 186 to 10,360 square kilometers. Socially, the effect was quite otherwise. As the British report to the League of Nations for 1927 put it, "The prospective pump-owner is usually an enterprising capitalist townsman, lacking land and anxious to develop a portion of the Domains already subject to tribal occupation." This "enterprising capitalist townsman" usually made a deal with one of the men the British had "promoted" to chieftainship, to convert tribal land to private property.

Alienation of the peasant from "his" land reached its logical conclusion in 1933. By that time, Iraq had become legally if superficially independent, with a parliament made up of the combination of urban investors and tribal leaders. They seized their opportunity in Law 28, "Governing the Rights and Duties of Cultivators." It effectively converted the formerly free tribesmen and villagers into serfs and solidified the dominance of the "chief" and "enterprising capitalist townsman." This stunning social conversion was effected by the definition of debt. So widely was it applied that no farmer was likely ever to be free of obligation: he owed not only for seed and equipment but for any work done by the landlord and for any assigned tasks he did not satisfactorily perform. He was tied to the land. If he tried to leave, the owner was authorized to call on the police or even the army to bring him back and make him pay for their services. If he managed to escape capture, he was blacklisted and prevented from finding further employment. Hatred generated by this social revolution would cause revolt after revolt until it exploded in the revolution of 1958.

• • •

When the mandate was devised, the British government negotiated a treaty with the men it had appointed as rulers of Iraq. This curious document was signed in 1922. While it reaffirmed the ultimate aim of Iraq achieving independence, it reserved to British authorities control of foreign affairs, the army, and finance. As we shall see, it was almost exactly copied by the "constitution" American authorities worked out with their Iraqi appointees in 2004. As Iraqis then felt and continued to feel throughout the period of "British Iraq" and later in the period of "American Iraq," what was termed an Iraqi government was, in practice, only a façade.

The treaty also created a legal absurdity: Britain retained an obligation as mandatory power to the League of Nations, but in Iraq it had replaced the mandate with a bilateral treaty. How to square this circle excited legal theorists. The answer was to proclaim a new constitution. In the mood of the times, when all over the world lawyers were writing elegant and sonorous constitutions, everyone agreed that the Iraqis should do the same. Months were spent discussing the proper words. Other constitutions were carefully studied to catch the proper cadence. Parts were borrowed even from distant lands like New Zealand. But the result was hollow. What emerged had little to do with the social or political realities of Iraq. And, when put to the test, as it was bound to be in the 1930s, the "constitution" proved to be merely a smudged scrap of paper.

Moreover, despite the sonorous phrases, the constitution that was finally adopted had been gutted of virtually all safeguards of political freedom. This was made clear in the first election held under British rule, in 1924. The handpicked members of both houses of parliament were not only chosen by indirect ballot but were carefully assisted to run unopposed. This electoral system—both the formal letter of the law and the means by which it was actually conducted—was retained until 1952, but in practice remained until the coup of 1958.

Again, a precedent for the future. One of the reasons that Iraqis reacted so sharply against the American-controlled "Iraq Provisional Authority" of 2004 was that in it they—but not the American authorities who were ignorant of Iraqi history—heard an echo of this early British system.

The importance of the British having inappropriately acted as though the unrepresentative and undemocratic government they installed was at least quasi-independent would become clear in the 1930s. Fronting for their British "advisers," forty Iraqis played a game of musical chairs in the twenty-one cabinets from 1921 to 1936; this system, so cleverly conceived, debased the very concept of representative government. If what the Iraqis knew of government under the mandate was "representative democracy," Iraqis were little inclined to support it. Democracy itself had become virtually a bad word. Thus, the Iraqis were set up to catch a mood or trend that also affected the much more sophisticated Europeans of the same time when various fascist movements came to the fore. In England, the man who had started it all in Iraq, Sir Arnold Wilson, joined the Fascist party.

In the 1930s the devaluation of representative government, lack of developed civic institutions, and the imbalance between the cities and the rural areas, the rich and the poor, the landed and the landless, the literate and the illiterate created a sense of frustration and anger. Perhaps even more important, it fostered the search for shortcuts. In that search, the army seemed to offer the most efficient, most modern, most available means of action. So, during the period from 1936 to 1941 it was always the arbiter of political life. Seven times in those years, it made or supported coups d'état. Particularly younger people, disgusted by what they regarded as greedy, corrupt, and incompetent politicians, men who had sold out to the British, thought—or hoped—that the army embodied a true national spirit. Through thick and (mostly) thin, they and their children would continue to believe this up to the present time. Indeed, that is the worst of all the legacies of "British Iraq."

To those who had conceived it, the concept of the mandate appeared quite otherwise. They portrayed it as a grand venture in education. The benighted peoples of Asia were to be taught to rule themselves. European "advisers" were to be their tutors, and they were to be their dutiful pupils. This too was to be echoed in the American approach to Iraq in 2003: We were to bring democracy to Iraq. Our willing students, the Iraqis would do as we instructed them, and their example would begin a great political revolution, a "crusade" in President Bush's unfortunate wording, that would transform the whole of Asia.

In Iraq, the British had virtually a clean slate. The Turks had done little to educate the Iraqis, and what little teaching was done in Iraq was in the Turkish language, which few Iraqis could understand. In the whole country, there were about fifty-four schools, all primary, which, on the eve of the war in 1913 enrolled, mostly only part-time, about six thousand children. In the traditional Islamic system, education was a community rather than a government responsibility, so a larger number of young people attended religious schools, Muslims in the mosques, Jews in their temples, and Christians in their churches.

At first, the British made few changes in the Turkish system except that they encouraged the use of Arabic and English. In 1919 the occupation government operated twenty-one schools.

All were primary schools in which only about half of one percent of the school age Iraqis were registered. Even fewer actually attended. In 1920 the British also opened two secondary schools, one in Baghdad (with twenty-seven students, of whom eighteen were Muslim) and one in Mosul (with seven students). In 1921 education took up 3.03 percent of the state budget. Religious schools remained predominant. In 1923 some three hundred Quranic schools enrolled fifteen thousand pupils. An additional five thousand adults attended anti-illiteracy classes of an organization known as Mahad al-Ilmi. But the British realized that they needed to justify their assumed role as "teachers" of nationhood and so they moved to create a rudimentary educational system.

By any definition, it was certainly rudimentary. By 1925 the mandate government was employing eight hundred teachers, but of these almost half had no formal education. Almost all these men worked in the major towns. In the smaller towns, a student was likely to get only two years of classwork, which, as a later educational mission from America pointed out, is below the critical point at which he will remember what he has learned and can build upon it. The nine in ten Iraqis who were then "rural" never saw a teacher. As the official report for 1923–1924 put it, "in this country, it is neither desirable nor practicable to provide secondary education except for the select few." The four secondary schools by then in operation, the report continued, were probably too many. In that spirit, little money or effort was spent to teach or train Iraqis; by the end of the mandate in 1932, education still got only a third of the amount allocated to the police.

Both teachers and students found what they were allowed to study dull and uninspiring. Issues and political ideas and programs that were regarded as subversive were, of course, off limits. Frustrated in schools, many enrolled in political movements, of which the most significant was an adaptation of the European youth movements. Borrowing a name from the Abbasid times, the Futuwah sought to instill nationalist ideals. Under the guidance of the director general of education, it became a native fascist movement. As in Europe during the same era, the failure of democracy to put meat on the bone of its ideals left many in Iraq hungry for what seemed powerful, modern, and purposeful.

Meanwhile, the privileged elite had realized that their sons needed more than could be provided inside Iraq, so increasingly they sent their sons abroad to study. The first group, of just nine young men, actually went in 1921. Evincing the distrust felt toward Britain, most went to American institutions. In 1923, for example, four went to England, two to America and 120 to the American University of Beirut. By the eve of the Second World War, the number abroad had reached 238. As the students began to return after completing their studies, from 1926 onward, they were quickly absorbed into the secondary schools and teacher training programs. However, many students found that the experience abroad had not so much trained them as alienated them. Most later reported how hard it was for them to find a meaningful outlet for their new skills. The major exception was the medical college, which was set up in 1927 and attached to Baghdad's hospital. Ironically, for a country with the commitment to agriculture that Iraq had, the school of agriculture was closed in 1930 for lack of students.

What was perhaps more important than the school system was the informal effect of commerce, casual contacts with foreigners, radio, the cinema, and the press. The new oil industry became virtually a separate state, and although it preferred Indian clerks and English managers, some Iraqis benefited from on-the-job training. Shops along the main street of Baghdad were apt to be staffed by Indians, but Iraqis, at first doing only menial tasks, inevitably handled new tools, used new products, and learned new ideas. Very slowly and with little help from government, Iraqis began to acquire the "tool kit" of a modern society; what they lacked was the balancing appreciation of civil life—a commitment to mutual respect under law and to peaceful participation in political institutions. For this imbalance they would pay a fearful price. The imbalance remains and undoubtedly will continue to degrade the quality of civil life.

What is particularly striking, I believe, is what a small role religion played in the formation of an Iraqi nationality. This is in sharp contrast to Egypt, where religion had helped to define "Egyptianness" and to lead the reaction against the British invaders. True, the Iraqi Muslim army took as its major nationalist task the destruction of the power of the Christian Assyrian levies. However, the record and numerous discussions I have had with contemporary Iraqis strongly suggest that the levies were hated not as Christians but as tools of the British. Those still under arms would prove this charge when they fought for the British during the brief battle that overthrew the Iraqi government in 1941.

Also, it was true during the years of "British Iraq" that the Shia "clergy" energized protest movements against the government, but it seems to me that they assumed this role less as a religious organization than as the only nongovernmental organization. Before the British co-opted the Sunnis, they and the Shiis had worked together against the British; after the Sunnis became associated with the government, the Shiis found themselves isolated in their opposition. Thus what was essentially a nationalist sentiment came to be cast partially in religious terms. And having taken on that guise, it acquired an organizational reality. It retains that reality—and for the same reasons—today.

• • •

In 1932 the British agreed to end their mandate, and Iraq signi-
fied its new formal status by joining the League of Nations. Just
a year later, the man the British had brought to Iraq as king
departed for medical treatment. Never loved by Iraqis, King
Faisal died in Switzerland and was followed by his much more
popular (because more nationalistic) son, Ghazi. Experimenting
with new ways to circumvent the British and reach the Iraqis,
Ghazi created a radio station to appeal to the still largely illiterate public and favored the intervention of the army increasingly
in civil affairs.

The army was also beginning to experiment. In 1936, after
having suppressed and massacred much of the Assyrian community, by then totally identified with British rule, and haring put
down a tribal rebellion, the army staged what was to be the first
of many coups d'état. From then on it was the arbiter of politics.
But the involvement of the king in its activities brought him
increasingly into conflict with the British, who, by then slightly
further behind the scenes, still managed Iraqi affairs. In 1939 the
king was killed in what was said to have been an automobile
crash but which Iraqis generally believed to have been a British-
instigated murder. Demonstrations broke out, and a British con-
sul was murdered in Mosul. Ghazi's infant son was named king
under the control of a strongly unpopular and strongly pro-
British regent.

Hostility toward Britain was general and was fanned by
British actions. Leading the opposition to Britain was the then
prime minister, Rashid Ali, and speaking on behalf of Britain
was its ambassador to Iraq. As relations deteriorated, the ambassador called in the foreign minister, Nuri Said, a man with
whom Britain's relations had always been close and who was to
play a major role in all subsequent events until 1958, and told

him that "Iraq had to choose between her friendship with Great Britain and her Prime Minister." Rashid Ali attempted to compromise, acceding even to a British demand to be allowed to land troops where they had begun, in Basra, but hatred of the British ran so deep that even the parliament, long since bought and paid for by British policy, unanimously moved to remove Britain's principal supporter, the regent, and replace him with his cousin.

As the Second World War broke out, the American government, for the first time, involved itself in Iraqi affairs. The American minister delivered a note "advising" the Iraqi government to "cooperate" with the British, saying that America intended to do all it could, "short of a declaration of war" to assist Britain. Iraqis, naturally concerned with their own affairs, took this to be a sign that their quasi-colonial status would be continued. Iraqi nationalists added America to their enemies list. As the gulf between the two sides widened, Britain cut arms supplies and the subsidy it had been giving the Iraqi army. Those acts provoked a group of senior officers to stage what amounted to a coup against the British-supported monarchy and in support of Prime Minister Rashid Ali.

The pro-British regent, with the help of the American legation, fled the country. He was quickly followed by the group of politicians and former army officers who had always been pro-British. This appeared to clear the field for those who had long agitated against the British, and they quickly focused their anger on the symbol of British power, the airbase the RAF maintained fifty miles from Baghdad.

The Iraqis regarded this airbase as a dagger aimed at their capital and, exaggerating their abilities, decided to close it down. Sending their still minuscule army to the area of the base, they ordered that all flying there be stopped. The British refused.

Regarding the Iraqi military with low esteem and determined to maintain their position, the British attacked the Iraqi army on May 2, 1941.

Still hoping for a compromise, Rashid Ali sent the British a conciliatory message but the Shia community leaders, the *mujtahids*, took an uncharacteristic leadership position and called for a holy war (Arabic: *jihad*) against the British. Their call was echoed by the grand mufti of Jerusalem, a Sunni, who had fled British-controlled Palestine. Rashid Ali cast about for foreign assistance against the British. America would not help. France was defeated and under virtual German control. That left the Germans and the Russians. The Germans initially turned him down, but the Soviet Union made a token gesture of diplomatic support that, under the circumstances, only made the British more determined to overthrow him. Belatedly the Germans saw some advantage in embarrassing the British, so they arranged that the French, then under the Vichy regime, send some guns and ammunition by train to the Iraqi army through neutral Turkey, and also sent a Luftwaffe squadron through Syria to Iraq. In response to these moves, the American government, itself still officially neutral, seized French merchant ships in American ports while the British prepared to invade French-mandated Syria and Lebanon. At the same time, the British also built up their forces at their airbase for an attack on Baghdad. With some fifteen hundred Assyrians, Indian troops, some Jordan Legionnaires, and a small English contingent, they defeated the Iraqi army in a battle at the Shia suburb of Baghdad, Kadhamain, on May 29, 1941. Prime Minister Rashid Ali fled first to Iran and then to "the enemy of his enemy," Germany; the senior Iraqi army officers were captured and hanged, and several hundred other officers were sent to internment camps.

The regent, the little boy king, and Nuri Said were returned to Iraq to be upheld by British steel. No Iraqi could doubt that "his" government was British. The war years were a hiatus for Iraq. All political activity ceased under British armed occupation.

As the war ended, some political activity had to be tolerated, and new groupings, not quite on the scale of political parties, emerged that were all anti-British. Joining the more traditional groups was a newly organized communist party; it was revolutionary not only in program but in membership. Both Jews and Christians figured prominently. For the first time, a political group began to address workers. And they listened.

Although violently suppressed, workers began to strike for better living conditions. A few small strikes had already occurred during the war, but as the war ended, both the scale and the dispersion of their activities accelerated. A 1945 strike by Baghdad railroad workers lasted more than a week and spread to Basra and other cities. The British management, backed by the government, tried to break the strike by threatening to cut off the water supply to the cantonments in which the Iraqi workers lived and to replace them with Indian workers. Since nothing was done about their very real grievances, the workers struck repeatedly in the next three years.

What the railway workers had begun was picked up in the only major industry Iraq had, oil. For the first time, the workers at the huge Kirkuk installation went on strike; a few were shot down by the police and many more were injured. Since the field was run by the British and the police operated under their control, the opposition saw labor unrest as nationalist: battle was joined. This battle was known in Iraq as the "aggression," (Arabic: *al-wathba*).

The British and the ruling establishment regarded *al-wathba* as a communist-led uprising. They were determined to crush it. Cleverly, if not wisely, Prime Minister Nuri Said decided that one way to weaken it was to remove the symbol that unified the disparate groups making up domestic opposition: British domination. Nuri worked with the regent to urge Britain to conclude yet another treaty in which the British relationship could be portrayed in a more favorable light. The Portsmouth treaty that emerged from lengthy but secret negotiations made some concessions to the nationalists, particularly in turning over to Iraq the British military bases, as Rashid Ali had earlier demanded, but specified a continuation of British supervision far into the future. When word of its nature leaked out, the students in government schools took to the streets. The government cracked down on them as it had on oil workers: police fired into the crowds. This harsh action broadened the opposition far beyond young students to include the emerging professional elite. Alarmed, the government struck at all its critics: further peaceful demonstrations were bloodily suppressed; the leaders of the communist party were arrested, tried, and publicly hanged; and scores of those who had actively agitated were arrested, tortured, and imprisoned or exiled.

It was at this point that the issue of Palestine boiled to the front of Iraqi consciousness. To the Iraqis, it then seemed that what the British had done in their mandate for Palestine was similar to what they did in Iraq. In Palestine, they had not only dominated the native Arabs but had given away their land to other Europeans for reasons that had nothing to do with the hopes and desires of the natives. To the Iraqi generation that had grown to adulthood at the end of the Second World War, defending the Palestinians came to seem the ultimate test of Arab "brotherhood."

The Iraqi part of the Arab failure in the 1948–1949 Arab-Israeli war seemed to the politically active younger people of Iraq not only a defeat but a national shame. Their soldiers had been marched off to fight in that war with little training; many did not even have guns or ammunition; some did not even have shoes or jackets. The Iraqis felt betrayed by their government; they were humiliated. Iraqi culture, like Arabic culture in general, is permeated with concern for shame (Arabic: *ar*). Palestine cut to their deepest emotions. And they believed the source of their sense of shame was precisely the British-dominated clique that had ruled the country with few interruptions since the First World War. At the peak of that system was the dour figure of the man who alternated between being prime minister and puppet master, Nuri Said.

Nuri was well aware of the hostility toward him, his government, and the monarchy. To protect the regime, he did three things. He cracked down on opponents, killing some, imprisoning others, and buying off many. He neutralized the army by ensuring that while on duty in Iraq, military units were deprived of ammunition and usually kept far from centers of power. And, at the same time, he began a program aimed at restructuring the country economically. In this effort, he was able to draw on substantially increased oil revenues.

Oil in commercial quantities began to be produced in the Kirkuk area in 1927. Since the status of the area was still in dispute—the British having seized it from the Turks in violation of their ceasefire at the end of the First World War—the mandate government compensated Turkey by paying a royalty of 10 percent for a period of twenty-five years.

Oil was not then so lucrative a commodity as it became when the Western world took to the automobile and the airplane, but in the first twenty years of Kirkuk production, the field yielded more than 100 million tons of oil. The Iraqi share of the proceeds was relatively small. Few Iraqis were initially employed, most of what the industry required was imported, and the small payments made to the Iraq government went mainly into the regular government budget. But during the

1950s new fields were opened, production grew, and the Iraqi share of profits was substantially increased. On the eve of the 1958 revolution, Iraq was receiving about $250 million from the British-dominated Iraq Petroleum Company (IPC).

As younger Iraqis were becoming better educated and were increasingly in contact with European and American sources of information, they began to be aware of the enormous importance of oil to their future. They came to believe that in oil policy as in other affairs, their government was corrupt or even traitorous. It had allowed the IPC to keep the lion's share of the profits. Thus, as their share increased, they were not so much grateful as critical of its having accepted the "jackal's share" for so many years.

The government sought to deflect such criticism by announcing in 1950 that it would allocate 70 percent of the oil revenues to a fund administered by a new quasi-independent "Development Board." With this greatly enhanced revenue—up threefold between 1951 and 1956—the board commissioned plans from European and American consultants for new highways, dams, canals, bridges, and other infrastructure projects. Plans, literally by the ton, began to accumulate in its offices.

In some fields, real and lasting accomplishments were made. Huge new dams were constructed to tame the rivers, highways and bridges were constructed. The board took as its major task the uplifting of Iraq's main domestic economic activity, agriculture. However, the ways in which it carried out its tasks actually exacerbated rather than solved existing social and economic problems. What it did was to promote schemes that brought new lands into production rather than attacking the "structural" problems of the existing lands. The new lands were quickly gobbled up by the combination of wealthy city men and "chiefs" who dominated parliament. They owned almost three-quarters of the land then suitable for agriculture, while poor farmers were left with lands that were far less productive.

Consequently, despite the relatively vast new investments, agricultural production fell. The World Bank and other agencies saw the problem and advised against land reclamation, but the Development Board gave in to the ruling oligarchy and continued the course upon which it had embarked. In the name of reclaiming the Garden of Eden, it created the conditions for the overthrow of the government in 1958.

Worse, little was done to increase the capacity of the people: illiteracy was at such a high level that many of the younger people in Baghdad believed keeping Iraqis ignorant was government policy.

Objectively speaking, the accomplishments of these years were significant, but in the estimation of the growing number of young men and women who were returning from study abroad armed with the skills needed to implement the plans and with heightened sensitivity to the politics of Arab nationalism, these positive moves were devalued. The government responded to their criticism by imposing martial law. It was just at this point that young Iraqis were electrified by the Egyptian coup d'état of July 1952. Weighed in the scale offered by Gamal Abdul Nasser, even the best that was done in Iraq was found wanting. Each inch forward was not regarded as an improvement so much as the failure not to have moved a yard. And, politically, many felt that movement was not toward a brilliant future but just another step along a path chosen in 1920, toward Iraq as an outpost of the British Empire.

It was at this point in 1955 that Iraq became the centerpiece for the Anglo-American Baghdad Pact. John Foster Dulles's fateful demand that the Iraqis "stand to be counted" in the Cold War against the Soviet Union was the move that set the stage for the 1958 coup d'état that ushered in "Revolutionary Iraq."

FOUR

REVOLUTIONARY IRAQ

For half a century, since the last days of the Ottoman Empire, young Iraqi army officers gathered to discuss politics. They convinced themselves that they had a "sacred" mission to protect their nation because they believed that they were the only ones who were above the corruption so evident among civilian politicians. They also proudly observed that they were the only ones who had the power to effect the nation's will. Their fathers or grandfathers had formed secret societies before the First World War, had helped to bring about the Arab Revolt, and had overturned governments they regarded as traitors during the years of "British Iraq." The young men followed their tradition.

As their plots were revealed and coups thwarted, group after group had been exiled, imprisoned, or shot. However, since even civilian governments thought Iraq needed an army, an unending stream of recruits reconstituted their cadres. Always the disaffected found brother officers who were willing to listen to them and soldiers who were willing to obey their orders. Nationalism was their cause. Its growth among the Arabs had been both recent and indirect. They had lived for centuries under the Ottoman Empire, which was composed of a variety of ethnic groups that it sought to accommodate. Then, in the early years of the nineteenth century, various of its component nations began to assert their separate political identities. First in the Balkans, the Greeks, Bulgarians, and Romanians led the way; then in Anatolia, the empire's central province, the Armenians followed. These disaffections finally caused the Turks to probe their own identity as something separate from "Ottoman." Their ideas coalesced in a vague sense of "Turkishness" (Turkish: *Türkçülük*) which found expression in the "Young Turk Revolution" of 1908. When its leaders began to assert their new sense of national identity, their action forced young Arab army officers, serving in the Ottoman army, to search for an identity that was not Turkish. What could be this Arab "nation" from which identity derived, to which loyalty was owed, and in whose cause men were soon willing to risk death?

Unlike the other nations of the Ottoman Empire, the Arabs had no self-evident answer: whereas the Greeks, Bulgarians, Romanians, Armenians, and even Turks were unified by language, religion, way of life, and geography, the Arabs were diverse in each of these. True, their literary language was uniform, but their everyday language was divided into scores of dialects of which some were mutually unintelligible. In religion, they were even more diverse. Some were Christian, and among Muslims, distinctions between the Sunnis and the Shiis were not only liturgically significant but were antagonistic historical experiences. In manner of living, the division between settled (Arabic: *hadar*) and nomad (Arabic: *badu*) had always made Arabs seem alien to one another. And, finally, the vast stretch of geography imposed upon them historical experiences so diverse as to be almost incomprehensible each to the others. So, as they began to try to come to grips with the question of identity, the Arabs had a particularly difficult set of obstacles to overcome. They have yet to do so to their common satisfaction.

Early attempts to overcome these obstacles occurred far from Iraq. First in Egypt and then in Lebanon, scholars began to rediscover their literary heritage. But they influenced few people. Then, in 1905, a young Christian Syrian published an anti-Turkish manifesto called *Le Réveil de la Nation Arabe*. The fact that the manifesto was in French and by a Christian indicates the problem Arab thinkers had. Since the problem of definition, thus announced, would echo through Arab politics for the following century, it is crucial to understanding future events.

First, consider religion. As we have seen, during the Arab conquest of what became Iraq, the invading Muslims wanted the natives to remain non-Arab. They treated the Arabized converts (Arabic: *muwali*) as outsiders. It would be centuries before the converts became full members of the dominant society. Christians and Jews took even longer. In Iraq, it was more or less accomplished in the 1920s, as symbolized by the membership of an Iraqi Jew in the first cabinet in the government of "British Iraq" and the rise to prominence of both Christians and Jews in commerce, education, and administration. Then acceptance of them as fellow members of the new nation was set back by events over which Iraqis had no control. The position of Christians was undermined by the British use of Christians, the Assyrian levies, to maintain their domination of Iraq, and the rise of Zionism, the Israeli-Arab wars, and the plight of the Palestinian refugees made the position of Jews ultimately untenable.

Second, geography. The experiences of Arabic-speaking people in the several countries were very different. The 1905 manifesto was written by a man living in the Ottoman province of Damascus, that is, in what became Syria. Syria and adjoining Lebanon had long been regarded by the French as at least culturally theirs; in 1920 France conquered them. That had the dual effect of making many Syrians hate France while putting them under French cultural influence. Egypt had been conquered by Napoleon in 1798 but the French were thrown out after a few years. What remained and grew during the century was a respect for French culture. It was England, which occupied Egypt in 1882 and ruled it directly or indirectly thereafter until 1952, that was the imperial enemy. Palestine, Jordan, and Iraq were invaded and ruled by Great Britain from the middle of the First World War. Libya was conquered and ruled by Italy from shortly before the First World War until the middle of the Second World War. Consequently, each of these areas received different outside influences: the natives had different tutors, rulers, *and* enemies. As I observed when in 1952 I attended a conference of scholars from the various areas, they had considerable difficulty in conveying their ideas to one another in Arabic. Having studied in Europe or under Europeans, some thought in English, others in French, and still others in Italian or German. What was true of intellectuals, of course, was even more true of craftsmen, merchants, and soldiers. Each mandate or colony took on aspects of a separate nation. The Arabic word for this geographical nation is *watan*.

Some leaders of the Arab countries accepted this reality and cited it to justify their roles. In Iraq, while members of the cabinets of "British Iraq" paid lip service to broad concepts of an Arab nation, they focused on Iraq as a state. This was the policy of the perennial prime minister, Nuri Said; it also became the strategy of the man who overthrew him, the first leader of "Revolutionary Iraq," General Abdul Karim Qasim, and of some of his successors, including Saddam Husain. Even when ostensibly struggling against "particularist nationalism" (Arabic: *wataniyah*), they would be driven to act in accord with it.

Many politically active army officers and civilians in Egypt, Syria, and elsewhere, were infuriated that their European conquerors had so divided them; for them, nationalism based on the *watan* was a perversion hatched by imperialism. The fact that it was also a reality only made it more odious. True nationalism, they proclaimed, was based on the people; for that idea of nationalism the Arabs adapted an evocative term derived from the group to which loyalty and identity were absolute, the clan. Unconsciously, they followed in Muhammad's footsteps when he sought to overcome tribal divisions by treating early Islamic society as a sort of clan. Since they were primarily secular, they altered the definition of the clan (Arabic: *qawm*) from religion to ethnicity. Thus "folk" or ethnic nationalism is known in Arabic as *qawmiyah*. *Qawmiyah* became the matrix of the politically active young Iraqis, of their hero, Egyptian president Gamal Abdul Nasser, and, at least initially, of the "Awakening" or "Renaissance" (Arabic: Baath) movements in Syria and Iraq. For them all, true nationalism, "Arabness" (Arabic: *Arabiyah*) arose from a feeling parallel to the *Türkçülük* that had earlier driven them apart. Divided and driven by these two different definitions of nationalism, *wataniyah* and *qawmiyah*, Iraqi politicians and army officers would struggle for leadership between 1958 and 2003 in what I have called here "Revolutionary Iraq."

Both in office as prime minister and behind the scenes working through protégés, Nuri Said dominated Iraqi politics from the overthrow of Rashid Ali in 1941 to the revolution of 1958. While he had begun his career as an army officer fighting for Arab unity and had returned to Iraq only at the end of the First World War as one of King Faisal's party of young officers fresh from the "Revolt in the Desert," his concern had long been Iraq. Iraq was the seat of his personal power and the reason for his alliance with the British; it was also a political and economic entity that he thought was sustainable. Personal position and geographic reality dictated *wataniyah*: he was Iraq's leader, and Iraq was simply different from other Arab countries.

Iraq's security was affected by neighboring Iran and Turkey, with which distant Egypt, for example, had no serious concern. Also, while Egypt might flirt with the Soviet Union from its relative isolation, in Nuri's estimation the Soviet Union posed a serious threat to Iraq. Thus, from 1954, he began to move toward closer relations with Turkey and Iran; then in 1955 he severed relations with the Russians. When the United States, assuming Britain's role as the hegemon of the Middle East, began to put together what became known as the Baghdad Pact (officially CENTO), he eagerly joined. Again, history had repeated itself: as foreign minister twenty years before, Nuri had laid the groundwork for a comparable (but mainly anti-Kurdish rather than anti-Communist) alliance known as the Saadabad Pact. For him, notions of pan-Arabism seemed a distraction, a weakening of Iraq's independence; more personally, it implied his taking second place to Egyptian president Gamal Abdul Nasser. Nuri was a confirmed practitioner of *wataniyah*.

To a growing group of younger Iraqi army officers, the survivors of frequent purges, Nuri's position seemed not only heresy but treason. They believed in *qawmiyah*. How to win them over or prevent their intrusion into politics was Nuri's ever-present problem. He went at it in two ways. On the one hand, he used the increasing revenues from oil to promote development schemes that particularly benefited the groups from which the officers came and catered to their demands by encouraging the United States to supply military equipment on a scale approaching the NATO standards. Despite his best efforts, plots continued to be hatched.

Usually Nuri managed to stifle dissent or purge those who

organized against him. But sometimes he could not. When a plot was discovered in 1956, his hands were tied by actions over which he had no control. The invasion of Egypt in that year by a consortium of Britain, France, and Israel came to seem an indictment of him because of his long association with Britain. He had to cater to pan-Arab, *qawmiyah* feelings so he protested to Britain, broke relations with France, and attempted to turn anti-Israeli feeling against the Iraqi communists and other left-ists. But when civilian demonstrations broke out not only in Sunni Baghdad and Mosul but also in Shia Najaf, he unleashed his police to beat down overt opposition. Behind closed doors, however, his repression escalated opposition and focused it on a new force that had recently entered Iraqi politics from Syria, the Baath.

The Baath had been founded a decade earlier in Damascus by a group of Syrians as a discussion group and had grown into a small but vigorous political party. Authoritarian, somewhat mystical, vaguely socialistic but determinedly pan-Arabist (as I learned in long discussions with its leaders). its ideas were intro-duced into Iraq in 1951 by a young Iraqi engineer by the name of Fuad ar-Rikabi, who would later be murdered by Saddam Husain. What was interesting about the early Baath was not its ideas but the concept of membership: Rikabi and his associates cast their net wider than just army officers to include the grow-ing number of educated professionals. The issues they raised were more exciting than the usual discussion of the composition of cabinets; they focused not only on *qawmiyah* but also on gross social iniquities that grew out of the British "Tribal Disputes Regulations" and the Iraqi parliamentary Law 28 of 1932, "Governing the Rights and Duties of Cultivators." Few civilians joined Rikabi, but he did influence several small cliques of army officers. One was to come to the fore in 1958: it was made up of less than a dozen "Free Officers" but was significant because they actually commanded military units. Their influence spread when they invited a more senior officer, Brigadier General Abdul Karim Qasim, to become their leader. Qasim, in turn, brought in another group led by a protégé, Colonel Abdus-Salam Arif, who was already deeply wedded to pan-Arabism.

These loosely associated and vaguely allied officers were electrified in February 1958 by what looked like a major step toward Arab unity, Gamal Abdul Nasser's formation of the "United Arab Republic." To this challenge, as he perceived it, Nuri reacted by bringing about the formation of a rival union of Iraq and Jordan to which, it was presumed, Kuwait would adhere. What turned out to be important about this new union was the reverse of what Nuri had planned: when he sought to bolster the union by deploying units of the Iraq army to the Jordan frontier, he made possible a coup d'état. To get to the frontier, the army had to march past Baghdad. That was the opportunity their commanders were waiting for.

On the night of July 13, the force under the command of Colonel Arif seized what Lenin had once called the "heights of power." For Baghdad, they included the radio station and the royal palace. Taking over the radio, Arif announced the end of the monarchy, and at the palace his men assured it by killing the king and his uncle, the éminence grise of the regime, Prince Abdul-Ilah. They narrowly missed Nuri, who escaped across the river, as he had done once before in 1941. After a furtive chase, he was gunned down in a Baghdad street. With these few shots, the old regime collapsed. The coup leaders were stunned by how easy it had been, while in the streets astonished crowds danced to what they took to be the music of a new era.

Almost immediately, both groups had to face a rather different reality. The new government found, as I discovered in long conversations with its members a few days later, that while the symbols of the old regime were gone, much of its substance lingered. Many of the men and women I had known five years before had become ministers. They thought that the coup was made in their name. Most were soon to be not only imprisoned, but exiled or shot. In Iraq as elsewhere, revolutions have a habit of consuming their own. The military masters continued to talk of Arab brotherhood, *qawmiyah*, but they quickly found that running Iraq required them to use some of the methods Nuri had employed. The state was in their hands, and its resources were at their disposal. Moreover, they, and, particularly Abdul Karim Qasim, had tasted the "apple" of power and were no longer, if they had ever been, so keen on handing it to the immensely popular pan-Arabist leader Gamal Abdul Nasser. Whatever Qasim's personal feelings at that early stage, he was urged by various groups, particularly the communists and their allies, whose Egyptian counterparts Nasser had roughly suppressed, that a close relationship with Egypt would bring Iraq and him personally only danger. The line he soon adopted was "Iraq First," *wataniyah*. For him soon there was no second, no *qawmiyah*.

It was a few months after this time when I first met Qasim. I had returned to Baghdad to assist the German-American architect Walter Gropius renegotiate the proposal he had submitted to Nuri Said to design the campus for a university. To beat down the fee, Qasim had stage-managed the negotiation by pretending to televise the meeting live. The implication was that since his prestige as prime minister was engaged publicly, Gropius would have to acquiesce. The fee was large, so I suggested reducing it by a half percent. Qasim waved that aside, saying, "I do not talk in half percentages." By then I had guessed that the klieg lights

and whirling cameras were only a ploy, so I shrugged and suggested we go back to the original offer. Qasim burst out laugh-
ing and waved the cameramen away. It had been a game, and he
thought it rather fun. We got up and shook hands. As I was
leaving, I realized that I had left my briefcase behind; so I turned
around, climbed the stairs to his office, and walked in. There
were no guards about. I was surprised to see him sitting, feet up
on a desk with his collar unbuttoned and his coat thrown over a
chair. He smiled, got up, and pointed toward my briefcase. "Is
that what you forgot?" I nodded. "Well," he laughed, "you certainly didn't forget that other thing! [the fee]"

Qasim did not long survive the sycophants who praised his
brilliance. The voice he heard was the one that urged him to
assume one of those titles that echo down Iraqi history from the
first *lugal*: a man of modest attainments, he drank deeply from
the cup of vanity to glory in being called "Sole Leader." When
his comrade-in-arms, Colonel Arif, began to gain an independent following, Qasim packed him off in polite exile as ambassador to Germany. When he tried to return, Qasim had him
arrested. Those who hoped that the revolution had opened Iraq
to popular participation were quickly disappointed. As any
group began, or seemed to begin, to accumulate an independent
position, it became an enemy; even his supporters—Kurds, communists, fellow army officers—learned to dread his eye.

In these circumstances, former prime minister Rashid Ali
returned to Iraq from a long exile first in Iran, then in Germany,
and finally in Saudi Arabia. Apparently feeling that events were
replaying the 1941 struggle over Iraq's Arab soul, Rashid Ali
indiscreetly contacted some of those officers who were angry that
Qasim was following Nuri's *wataniyah* line. Qasim immediately
had him arrested, tried, and sentenced to death. Nuri would have
approved.

More serious were other challenges. In the spring of 1959, Mosul was plunged into what amounted to a civil war; alarmed, Qasim realized that he could not trust the army. It might be the problem rather than the solution. So he came increasingly to rely on those who seemed able to deliver popular "resistance." Best at mobilizing the "street" were the remaining members of the tiny Iraqi Communist Party. They, more than he, feared the Baath Party, whose members had wrapped themselves in the mantle of pan-Arab nationalism, *qawmiyah*. For their presumed role in the Mosul uprising, a number of their adherents were arrested, tried, and executed. That event provoked an attempt on Qasim's life in which a young, then unknown man from Tikrit, Saddam Husain, took part. Delightedly, Qasim showed me his bloodstained uniform in a glass case in his office. "They were not professionals, not serious," he said. "You always fire a second burst. They didn't. Too bad for them."

Useful as they were against the Baathists, he came also to distrust the communists; when they appeared to be gaining power, Qasim moved to weaken them. But, since he also allowed some of their leaders to remain in the government, he convinced some British and American intelligence officers, prone as they were to see communists under every bed, that he was a communist "tool." It was probably at this time that they too joined in the plots against him. Probably with help from them, those secular nationalists who had inspired the opposition to the monarchy and who remained out of jail began to conspire. Separately, proponents of political Islam began to become more active. The Sunnis had begun to create in Iraq a clone of the Muslim Brotherhood (Arabic: *Ikhwan al-Muslimin*) that had grown in Egypt in the 1930s and 1940s, but more important were the Shiis. It was at this period that they spawned the first of several parallel movements that would later oppose first Saddam

Husain and later the American occupation, the "Call" or "Prayer" (Arabic: *ad-Daawah*); by the end of 1960, it too had been added to Qasim's enemy list. Although fighting off rivals took most of his time and effort, at least in his early days, Qasim also promoted a number of social, educational, and health programs that were beyond even the dreams of reformers during the period of "British Iraq."

As they would always be, the Kurds remained outside the mainstream of Iraqi affairs. Toward them, Qasim had initially made conciliatory gestures, but, by 1961, the Iraqi army was engaged in a war against the longtime Kurdish leader Mulla Mustafa Barzani's "vanguard" (Kurdish: *peshmerga*) guerrillas. This little war drained the Iraqi state and increased criticism of Qasim in the officer corps, which realized that it was unwinnable. It also afforded an opportunity for outsiders—Americans, Israelis, and Persians—to weaken his rule, and they all used it discreetly but effectively by encouraging the Kurds in their resistance and supplying them the means to sustain it.

Qasim gave Iran cause for its covert war against him by claims on an adjacent province of Iran, Khuzistan, where the population was largely Arabic-speaking. Adding enemies as he went, he resurrected the Iraqi claim to Kuwait. Harking back to the Ottoman form, Qasim insultingly but popularly designated Kuwait's ruling shaikh "district governor" (Turkish: *qaimaqam*) of "that part of the Iraqi province of Basra known as Kuwait." That Kuwait was an integral part of Iraq, illegally split off by Britain from the Ottoman province of Basra and so a remnant of imperialism, was an old theme, first announced by King Faisal in the 1920s and picked up by his son, Ghazi, in the 1930s. There was truth to the Iraqi assertion, as many Kuwaitis acknowledged—the Kuwait Legislative Council had voted in the 1930s for union with Iraq—but by 1961, Kuwait had won, by a gener-

ous foreign aid program, general recognition for its independence. Even that pan-Arabist President Nasser sent Egyptian
troops there under an Arab League resolution to protect it
against Iraq.

In all these mostly maladroit ventures in both domestic and
foreign affairs, Qasim had indeed become, in a sense he did not
intend, a sole leader. His followers fell away in droves. Taking
advantage of his having alienated so many groups, the Baath
Party began to regroup. Working underground, it established
cells in the army, as Qasim himself had previously done, On
February 8, 1963, allegedly with assistance from the CIA[*] which
had been alarmed by Qasim's flirtation with communism, a
loose coalition of officers broke into the ministry of defense, captured Qasim, and, after a drumhead court martial, shot him.

The members of the coup, grandly designating themselves the
"National Council of the Revolutionary Command," installed
Qasim's erstwhile ally, Colonel Abdus-Salam Arif, recently
released from prison, as president, with a prominent Baathist,
Hasan al-Bakr, as prime minister. The Baathists thought that at last
they had arrived. To ensure their rule, they carried out a vicious
purge in which hundreds or perhaps thousands of members of
Qasim's regime were killed. Allegedly, again, the CIA helped them
identify those to be eliminated. Having killed off so many rivals,
they seemed secure, but almost immediately, the members of the

---

[*] Although I was the Member of the Policy Planning Council responsible for
Iraq at that time, I was cut out of information on these events. Whatever the
CIA did was presumably sanctioned by a special group, the so-called 40 com-
mittee.

council fell out with one another. The principal issue was the same that had haunted Iraq since Nuri Said's time and permeated Qasim's regime—should Iraq merge into a pan-Arab state (as Arif was demanding), that is, be guided by *qawmiyah*, or remain separate (as most of the Baathists were beginning to assert), as justified by *wataniyah*? Each side surreptitiously organized its forces so that, for a while, Iraq hovered on the brink of civil war. The conflict was resolved by yet another coup d'état, a coup within the previous coup: on November 18, 1963, Arif seized the power he had narrowly missed taking five years earlier on July 14, 1958.

Not surprisingly, given his clandestine background, his role as Qasim's understudy, his exile, his imprisonment, and his final seizure of power, Arif moved most easily in the shadows. A thorough authoritarian, he monopolized power, keeping an attentive personal control over the army. To block others from conspiring against him, he formed what later became the Republican Guard under the command of a close kinsman from his home district—moves that Saddam Husain would later copy. He quickly learned to play on both the theme of Arab unity (the intent to achieve which was built into the new constitution) and Iraqi separatism, which, though not publicized, could be seen in the way he used the resources of the state to build his political base. His final break with the pan-Arab faction came when its leaders attempted a coup d'état in September 1965. He had been wise to create his Republican Guard; they were the people who saved him. But what his enemies had failed to do was accomplished by what may have been an accident, a helicopter crash in the spring of 1966. He was immediately followed by his brother, Abdur-Rahman Arif, who continued his program.

Briefly, under the brothers Arif, it seemed that the Kurdish war, so long a running sore in Iraqi affairs, might be brought to a halt. The first Arif had appointed a civilian lawyer as prime minister, and among his programs was a serious offer to the Kurds of autonomy. This hope was dashed when the officer corps forced Abdur-Rahman to remove him. The war continued.

At this point, the Baathists seemed to be no longer a serious contender for power, since they had lost the support of the officer corps. But Abdur-Rahman's position had also deteriorated. Iraq's failure to assist other Arabs in the June 1967 war with Israel provoked widespread demonstrations. Unable to crush them, the regime seemed to have lost its vigor. As always in Iraq, that raised the possibility of a coup d'état. The trick was to neutralize Arif's Republican Guard. With the help of the United States, a small group of Baathists did this by cleverly playing on disappointments and grievances among officers who were not their adherents. Since no one was frightened of them, they were able to worm their ways into the few weak points in the regime's structure. But they could not reach the linchpin of Arif's regime, a kinsman and commander of his Republican Guard. Their opportunity came when he briefly went abroad. Then, just as Abdus-Salim Arif had done in 1958, so almost to the day a decade later, they seized the radio station, the ministry of defense, and the headquarters of the Republican Guard. A surprised Arif, who had thought the Baathists were a spent force and that they were isolated, agreed to resign, and departed for London. The coup had been bloodless.

That coup was just the first stage. In it the Baathists were only a part of a coalition. The second stage came two weeks later when they forced out their coalition partners and took sole power. A minority, they played out in 1968 on a small scale in Iraq what the Bolsheviks had done with their rivals in 1917 in Russia. Like the Bolsheviks, they then proceeded to use the power of the state to accomplish administratively what their small organization could not have done politically.

Also like the Bolsheviks in Russia, the Baathists in Iraq profited from the fact that the very concept of representative government seemed alien; worse, when known, it seemed an aspect, even a cause, of weakness, corruption, and disunity. It was this general lack of experience in and disregard for the workings of democracy, the legacy of both "British Iraq" and the first decade of "Revolutionary Iraq," that created the conditions in which the brutal dictatorship of Saddam Husain flourished.

In this new order, Saddam Husain played initially a minor, indeed almost a hidden, role. At the forefront of power was a man of evident power and prestige. General Hassan al-Bakr accumulated the posts of president, prime minister, secretary-general of the Baath Party, and chairman of the group known from Nasser's Egypt as the "Revolutionary Command Council." Saddam professed to be only his acolyte. Humble in public, especially before this man of the older generation, he moved steadily, determinedly and stealthily to create a new political reality that he would dominate because he alone knew what it was. The analogy of Hitler in 1932, deferential in front of the aged Field Marshal von Hindenburg while he built a new form of power, is perhaps not wholly incorrect.

Unlike Bakr, the Arifs, and Qasim—but like Hitler—Saddam was neither an army officer nor a professional man. Coming from a poor, rural background, he had little formal education. But driven from early adulthood to excel, he was a voracious reader; with little taste for the "overview," he delighted in detail; and with a phenomenal memory, he was also a tireless organizer. To draw upon another comparison, he could be compared to Vyacheslav Molotov, whom Josef Stalin had dubbed "the best file clerk in Moscow." Truly Molotov's worthy successor, Saddam made it his job to know the "name, rank, and serial number" of virtually every adult Iraqi. Knowing was only the first step. Using state patronage and appointment to official positions, he set out to align with what was then only a small cadre of Baathists virtually the entire adult Iraqi population. No petty civil servant, no teacher, no postman—one is tempted to say no street sweeper—was beneath his interest. All could be co-opted, bribed, or, if necessary, frightened into becoming his partisans. Through his efforts, even before he pushed General Bakr aside in 1979, the initially tiny Baath would grow into a mass political party, larger, proportional to the population, than Hitler's Nazis, Mussolini's Fascists, or Stalin's Communists. But the party and its ideology were only means. Saddam's single unifying aim was neither organization nor ideology—he wavered as danger or opportunity demanded between *qawmiyah* and *wataniyah*, between socialism and capitalism, between party and state, even between party and family. His aim was power. Whether he read Machiavelli or not, he would have agreed that the supreme interest of the ruler was power—how to get it, how to keep it, how to use it. These were the driving forces of his life.

Like other politicians who delve into the dark forces of humanity, Saddam realized that men need enemies. Much of his career would be spent finding them. Although he does not seem to have known much Iraqi history, he certainly knew that his contemporaries had been through so many coups that they imagined conspiracies everywhere they looked. He too feared them. But he also knew how to use them to justify his consolidation of power. He had many potential candidates—fellow Baathists, army officers, Kurds, Shiis, religious leaders, the British, the Americans, Persians, Syrians, Egyptians, and Israelis; eventually he would use them all. He began with the Shiis.

Sunni Muslims like Saddam and the men around him simply

could not believe that the Shiis, who had been generally excluded from participation in the Iraqi state since its inception, could be loyal citizens. Influenced by Persian culture, often speaking Persian and with relatives and friends in Iran, Shiis were regarded not only by the Sunni Arabs but even by the British as non-Iraqi. In Saddam's time, the Iranian government tarred them with its brush when it set up a radio station that urged them to overthrow the Iraqi state. Later, Iraqi religious leaders (Arabic: *mujtahids*, members of the *marjiaah* or "religious authority") ordered them not to join Saddam's party and issued rulings (Arabic: *fatwas*) condeming various of his programs. Old memories were thus augmented by current actions; by perverse logic, the fact that the Shiis fought bravely for Iraq during its war with Iran only made them seem more dangerous. They would be his regime's "enemy-in-potential" for the next forty years.

Treating them as disloyal tended, of course, to make them at least reserved—dissimulation (Arabic: *taqiyah*) was their traditional defense against tyranny—and this, in turn, made them seem more suspicious. Dissimulation was bad, but public protest was provocative. Whenever Shiis gathered, the security forces and the army moved brutally against them. Individual leaders, especially clerics, were arrested, imprisoned, or executed; their schools closed; their sermons silenced. The father and aunt of one of today's most important Shia leaders, Moktada as-Sadr, were among the thousands of victims. In 1969 the Iraqi government drove about twenty thousand of its Shia citizens across the border into Iran. When he could not defeat them, Saddam tried to join them. He made the birthday of the *imam* Ali a national holiday, rebuilt their mosques and places of pilgrimage, and even proclaimed that he was a descendant of the fourth caliph, Ali. All to no avail: to him, the Shiis were a different people. For that they paid in blood.

Even more alien and dangerous to the state than the Shiis, the Baathists believed, were the Kurds. Already in the creation of the state, the British had regarded the Kurdish issue as too difficult to handle. Had it not been for the discovery of oil in Kurdish-inhabited territory, near Kirkuk, and the danger that an unfriendly power might use a Kurdish state as a base, the British might have let them become independent. They did not, but they looked upon Kurdistan rather as they looked upon the Northwest Frontier Province of India—an unruly, tribal no-man's-land better left as much alone as possible. So it was that when Iraq became formally independent in the 1930s, Kurdistan was still "undigested." It would try, time after time, to force the Iraqi state to cough it up. Guerrilla warfare and ceasefire, hit-and-run raids and negotiations, acquiescence and defiance, participation in Iraqi government and conspiracy with foreign powers formed stages in an unending cycle. At the current stage of that cycle, it was met by massive military campaigns.

In one of those echoes of the long reach of history I have recounted, Saddam, almost certainly unknowingly, copied the policy of Iraq's remote ancestor, ancient Assyria, when it forced massive exchanges of population. In the 1970s the Baathists drove tens of thousands of Kurds across the frontier into Iran; when Iran and Iraq in 1975 worked out a deal to end their covert war, tens of thousands more fled. In that deal, the shah of Iran sold out the Kurds for a piece of the waterway leading to the Persian Gulf.* With their hands thus freed, the Iraqis began

---

* At the request of the shah, Saddam also expelled Ayatollah Ruhollah Khomeini, who had been living relatively quietly in exile in Najaf. The shah's request must rank as one of the more maladroit searches for national security in this century.

scorched-earth operations in which thousands of villages all along the Iranian frontier were destroyed. It then rounded up and moved Kurds into southern Iraq and lured or drafted Arabs to migrate to Kurdistan. These policies temporarily brought about a "peace of the exhausted," but they did not remove the challenges to Saddam's rule because Kurds and Shiis were not alone.

Since the inception of the state, Iraqis of all walks of life firmly believed in what might be called the James Bond school of politics. Behind every pronouncement, every alliance, every action, they believed, lurked ruthless, sinister, and brilliant foreign agents. Every Iraqi I ever met thought that British agents had murdered their King Ghazi; behind Nuri Said, they glimpsed shadowy figures at the British embassy; whenever Iraqis secretly met, they were sure that word immediately reached the British through cleverly placed agents. What the British did, or were thought to have done, became the heritage of the CIA. It was thought of as just the old MI-6 with more money. The Iraqis heard Americans boast that the CIA (with some help from MI-6) had overthrown the Iranian government of Prime Minister Mohammed Mossadeq, and they were convinced—rightly—that the CIA also played a role in the overthrow of Qasim. They also knew that the CIA worked closely with, and partially subsidized, the Israeli intelligence agency Mossad.

While there was much truth behind what outsiders usually dismissed as paranoia, these beliefs also revealed the basis of the insecurity of Iraqis. They realized that their country was small and weak but, possessing vast riches of oil, was a target for foreign domination. Iraqis knew that at least some of their earlier leaders had served as agents for foreign powers and supposed that current leaders were willing to do so. Knowing how high the stakes were, and how naked was the brutal struggle for power, they suspected, often correctly, that colleagues, friends, even relatives were willing to be co-opted. Thus a mood, somewhat like, but far more pervasive than the one that frightened Americans during the McCarthy period, prevailed in Iraq. Trust was a luxury; caution, a necessity.

In the hands of a master propagandist, as Saddam certainly was, there was enough reality to make almost any rumor or fear believable. Saddam acted as though enemies were everywhere. He used this to strike against actual or presumed rivals much as Stalin did in Russia. Thus it was that shortly after taking power, the Baath began to focus popular attention on a variety of groups and individuals, and the public generally credited the charges as true or prudently did not ask for proof.

Among the first victims was the popular former mayor of Baghdad who, under torture, was forced to confess to a plot. Could anyone be surprised? After years of conspiracies, Baghdadis joked that someone had actually found a man who was not involved in a plot. Foreigners were always suspect. High on the list was, of course, Israel. Who were Israeli intelligence "assets"? Naturally, the Iraqi Jewish community. Soon a Mossad plot was uncovered.[*] Whether it was true or not, it was logical; after all, Iraq was legally at war with Israel; enemies always

---

[*] Fourteen convicted prisoners, including nine Jews, were publicly hanged in the central square in Baghdad after a massive press exposé. The bodies were left hanging and crowds were urged to come to observe. So obscene did President Nasser regard this act that he personally called General Bakr, threatening to denounce him and his movement. Nasser's threat caused Bakr to stop the macabre demonstration. But Saddam had already made his point. The fate of those who opposed the revolution was made clear.

engage in intelligence gathering and espionage. Dozens of stories were told that illustrated how well informed Mossad was. If Israel was a prime suspect, it was only one of many: against both natives and foreigners, the formula of "discovery," arrest, show trial, and execution or "disappearance" would become as persistent and sordid a theme of the Iraq of the 1970s as of the Soviet Union in the 1930s. In this way, person by person, group by group, Saddam got rid of actual or potential rivals or frightened others away from temptation.

While he was certainly willing to use, and probably personally delighted in, ghastly forms of repression, Saddam recognized that fear alone would not ensure the survival of his regime. In such a tightly knit, "tribal" society as Iraq, to avoid vengeance, leaders had to cater to the imperatives of kinship; so while he mangled or murdered domestic and foreign real or presumed enemies, Saddam also moved to win favor. The means he used fell into three categories. First, he arranged to release the imprisoned enemies of previous regimes. For that they and their numerous relatives were grateful. Moreover, releasing some made the detention of others seem more reasonable. Second, he restored the jobs of thousands who had lost them either for political or financial reasons. He made sure they realized that they owed him personally their gratitude and that favors could also be withdrawn. Third, and, most important, he picked up the theme, first set out by Nuri and followed intermittently by Qasim and the Arif brothers, of national development. He began to promote the growth of a middle class and to create the most modern nation-state with the best educated, healthiest, longest-lived population in the Arab world.

Whereas under "British Iraq" the men and women who actually put their hands into the soil were reduced to virtual serfdom under the British "Tribal Disputes Regulations" and the Iraqi parliamentary Law 28 of 1932, "Governing the Rights and Duties of Cultivators," the incoming Baath regime in 1969 began a process of distributing the former tribal lands that had been taken over by city merchants and tribal "chiefs." Within a few years, nearly a quarter of a million farmers had received sufficient land to support a family.

Perhaps even more impressive was the opening of health facilities and schools free to the general public. Within a decade, school enrollment had been doubled. Whereas in 1920, thirty students were receiving secondary education—judged by the British as perhaps too many—in 1985 nearly one and a half million were. In every category, the numbers were almost equally impressive. When I lived there in the 1950s, Iraq's mechanical engineers, all five of them, could sit in my living room; by 1980, numbering in the thousands, they could not fit in a large auditorium. In 1951 about one in four children died in infancy. The number fell to European and American standards until the 1990s. As more doctors were trained, even with the growing population, the ratio went from one doctor for each seven thousand in 1951 to 1 for each eighteen hundred. As income rose, going up almost tenfold, so did life expectancy increase from an estimated forty to fifty-seven years. In short, the difference between the Iraq in which I had lived in the 1950s and the one I saw in the 1980s was truly astonishing. For this, in fairness, Saddam deserves much credit. The tool in his hand was oil.

Oil was a part of the legacy of "British Iraq." The British organized, invested in, and managed the consortium known as the

Iraq Petroleum Company (IPC)* which operated the giant field at Kirkuk in northern Iraq. Prudently, the IPC regulated extraction there to ensure a steady and profitable market, and since the managing partner, British Petroleum (BP), profited more from fields in Iran and elsewhere, it limited what it produced in Iraq and largely prevented further exploration. Over these matters, the Iraq government had little influence.

Spurred by the deal worked out by the Arabian American Oil Company (ARAMCO), which split profits fifty-fifty with Saudi Arabia, Nuri negotiated a similar deal with the IPC in 1952. The result in Iraq was to increase Iraq government revenues from about $40 million in 1952 to nearly $238 million on the eve of the 1958 coup. To his credit, Nuri arranged that 70 percent of this new money would be earmarked for development.

Having realized that there was much more to be got from oil, Qasim moved to change the government-consortium relationship by demanding that the IPC give up 90 percent of the territory in its concession; he promulgated a law to this effect in December 1961. He also announced that he intended to create a national oil company. Both of these moves met strong resistance from the IPC, which was accustomed to dealing with compliant Iraqi governments. Angered by the IPC, Qasim took part in the creation of a producers combine, the Organization of Petroleum Exporting Countries (OPEC), whose objective was to shift the balance of power in favor of the producers. He made little progress, however, because of a slowdown in demand for oil on the world market. After he was deposed in the coup, Qasim's

---

* The IPC was composed of five companies: British Petroleum (BP), Shell, Esso, Mobil, and Compagnie Française des Petroles, plus the 5 percent interest of the Gulbenkians, who had put the original deal together. It was run by BP.

moves were followed in 1964 when Abdus-Salam Arif, working under better market conditions, created INOC (the Iraq National Oil Company). His brother, Abdur-Rahman Arif, later introduced competitors from France and Russia to open new fields and market Iraqi oil. Through these measures, the revenue from oil mounted swiftly.

Then, after another two coups, Saddam began in 1971 to study the oil issue. As he soon realized, the previous negotiations with the IPC had missed the essential point: while the terms governing the sharing of revenue were important, they were a factor of the amount of the production. If the IPC could decide to cut production, as it did by 50 percent in the Kirkuk field in 1972, or to hold production steady in Iraq while it exploited cheaper oil in other areas, then Iraq could never be truly independent. Saddam set out to overturn that system. Having concluded that what Qasim and the Arif brothers had done merely complicated the issue, he struck at the heart of it: in June 1972, the Iraqi government nationalized the IPC.

Nationalization of the IPC was perhaps the most popular move Saddam ever made. It is difficult for foreigners, particularly modern Americans, to understand how bitter the Iraqis were about foreign domination. There were (and still are) many sore points in their recent history, but foreign control of their economy ranks high. The generation that grew to adulthood after the Second World War regarded oil, Iraq's single major national asset, as the symbol of and reason for British imperialism.

I will later return to this subjective political and psychological issue in discussing America's relationship with Iraq and projections into the future. Here, I emphasize that, objectively, the impact on Iraq of nationalization was dramatic and positive. From 1973, when sale of its oil yielded Iraq $1 billion, revenues rose in just two years to $8 billion. That was only the beginning.

By 1980, it had reached $26 billion. This huge increase of revenues made possible the virtual remaking of Iraqi society, the building of vast new infrastructure projects, and the modernization and expansion of the armed forces. A golden age seemed to have begun. There was something in the programs for everyone, although, naturally, some benefited more than others; who benefited and how much he benefited were carefully controlled. So a new class of friends, relatives, and supporters of the ruling elite arose. The aim was power, but there was no denying the beneficial aspects of the doling out of money. Schools, universities, hospitals, factories, theaters and museums proliferated; employment became so universal that a labor shortage developed, and army officers, so long accustomed to British castoffs, began to receive the best equipment then available. There was enough money in those early days for both "guns and butter." Then it all began to fall apart.

In September 1980 began what would become Iraq's "quagmire," eight years of war with Iran that wasted its resources, costing about $15 billion yearly, killing tens of thousands of its people, losing it nearly fifty thousand young men who became prisoners of war, and nearly bankrupting it. Fought along a 725-mile, or 1,169-kilometer, front, the war resembled the ruthless, static trench warfare of the western front in the First World War; it would become proportionally far more costly for Iraq, then a country of 14 million, than America's war in Vietnam.

In addition to the money Iraq had to borrow, nearly ten times its then yearly revenues, it had "opportunity costs" of at least $25 billion when its oil exports were partly shut down by Syria. More visible, I found when I visited Baghdad in 1981, was the "development cost." The ambitious social and economic plan under which Iraq was on the way to becoming the most advanced of the Arab states was also a casualty: buildings stopped in mid-construction, projects abandoned, talents unused, enthusiasms unstimulated.

Of course, the Iraqis had reason to go to war. Governments nearly always have; it is the rare statesman who understands the terrible costs and dangers of war and works to solve outstanding problems in other ways. Saddam was not one of these rare statesmen. On the contrary, he saw it as an opportunity. Iran seemed weak; its fundamentalist revolutionary government had ruthlessly purged the shah's American-trained officer corps, and it had lost its American source of arms and supplies. Conversely, Saddam had never felt stronger. Conflict with the Kurds was, at least temporarily, lulled; Iraq's army had been massively equipped; its treasury was full. He got at least tacit approval from his Arab "brothers"—Jordan let him use the port at Aqaba, which was out of range of Iranian attack, while Saudi Arabia and Kuwait agreed to supply money. Moreover, Iran was provocative: Ayatollah Khomeini was urging the Shiis of Iraq to revolt, even to kill him; was thought to have incited coup attempts in 1980, and had ordered his army to shell Iraqi cities along the frontier. So on September 22, 1980, Saddam ordered his air force and army into action. His overt aim seems to have been to recapture territory he had ceded to the shah in 1975 to get him to stop aiding the Kurds, but clearly, behind that objective, was a fundamental philosophical clash as profound as the American-Russian clash over capitalism versus communism: Saddam believed in the secular nation-state and Khomeini in the fundamentalist religious state. Perhaps even deeper, Saddam apparently thought of his war as refighting the struggle that went back to the first century of Islam, Arab versus Persian. Lest anyone forget the historical analogy, he used ancient battles as code names for his own operations. Finally, as shortsighted strategists always proclaim, as Saddam did to a prominent Egyptian journalist, it would be short and simple: the war would be over in three months.

When I visited Iraq in 1983, three months had stretched into three years. The Iraqis were proud of their war effort and arranged for me to tour the front. What I saw were surely the most bizarre battlefields in the history of warfare. I was picked up in an air-conditioned Mercedes-Benz by an Iraqi former ambassador and a senior Baath Party official and driven on double divided paved highways almost to the front lines before having to transfer to a jeep. Miles of trenched tank traps knifed through the desert guaranteeing protection against the nonexistent Persian armor; row after neat row of late-model Russian armored personnel carriers and partially dug-in tanks were protected by batteries of surface-to-air missiles. I was supposed to be impressed and I was. But what really struck me was that even the dugout of the captain commanding a company on the front line was air-conditioned and illuminated by television. (When I visited him, he was engrossed in an Egyptian soap opera.) Through field glasses, I watched Iranian soldiers just across no-man's land two kilometers away doing their laundry. The war had settled into a way of life.

Initially, the campaign had gone well, but it quickly became clear that the Iranians were not prepared to admit defeat or to negotiate. Even purged officers returned to action, some direct from jail, and the Iranian troops, like those who had fought for Ismail Shah against Selim the Grim, went into battle as "witnesses" or willing martyrs (Arabic: *mustashhids*)—the Muslim equivalents of Japanese kamikazes—proclaiming their faith as they met almost certain death. Then a sort of stalemate followed up to the time of my visit, after which the Iranians took the offensive. Iraqi casualties mounted (probably eventually by war's end reaching well over one hundred thousand, with at least double that many wounded), and tens of thousands of Iraqi soldiers were captured. For a while, it seemed likely that the Iranian army would break through the Iraqi defenses. Almost worse, Iraqi oil revenues plummeted after the Iranians destroyed the oil facilities at the Iraqi port of Fao and the Syrians, with whom the Iraqis had fallen out, choked off the oil pipeline to the Mediterranean.

As the fighting dragged on and Iraqi fortunes waned, a remarkable revolt occurred in Baghdad in June 1982: members of groups that Saddam had kept apart from one another and that he believed constituted his most loyal followers—senior Baath Party officials, army officers, and even close relatives of his—met behind his back to draft an offer to Iran to end the war. Saddam was saved when Khomeini rejected a ceasefire. Khomeini's action provided an opportunity for Saddam to strike back at his wavering followers. He allegedly personally killed one member of the group, a minister in his cabinet; the group's candidate to resume the presidency if Saddam were deposed, Hasan al-Bakr, the former president and the man Saddam had treated as a father, then conveniently and suspiciously died. Moving to reorganize the state and party organizations, Saddam forced them to reaffirm his policy. Iraq would fight on. He had gone back to the *watanist* tradition. Frightening or disposing of critics and potential rivals and forcing a new unity, he survived. What really saved him, however, was American help.

American help came in several forms. Fearing that the Iranians would break through Iraqi defenses, the United States began to give the Iraqis satellite images showing Iranian military dispositions; these detailed and timely photographs enabled the Iraqi army to deploy effectively, causing tens of thousands of Iranian casualties and turning the tide of battle. It also began to supply arms and lent or gave Iraq money and foodstuffs without which the Iraqi economy would have collapsed. Perhaps as important—probably more important to Saddam personally—were political and diplomatic moves. As a presidential envoy, the later secretary of defense, Donald Rumsfeld, flew to Baghdad in December 1983 to identify the American government with Saddam and the Iraqi cause.* A month later, the United States removed Iraq from the "terrorist list" and added Iran; soon it implemented a policy known as "Operation Staunch" to stop its allies from selling or giving Iran arms. Finally, the U.S. Navy deployed in the Persian Gulf. It did this ostensibly to protect tankers carrying oil from Kuwait, but what was known as "the tanker war" went far beyond what was required to accomplish that objective: the U.S. Navy destroyed the Iranian navy.

That the American government gave Saddam what was literally regime-saving military assistance and psychological support was later seen as embarrassing when America began to speak of "regime change." It appeared particularly so because its new policy was implemented after Saddam was known to have embarked on a program to develop nuclear weapons—his installation had been bombed by the Israeli air force in June 1981—and after he had

---

* Long denied and illegally hidden from Congress, although of course known to Saddam, documents on his two missions have been declassified and can be viewed at www.nsarchive.org. Rumsfeld now says he "cautioned" Saddam not to use poison gas, but there is no mention of this in the minutes of his meetings. He said the United States was eager for friendly business and governmental relationships. Meanwhile, President Reagan had instructed government officials to do whatever was "necessary and legal" to prevent Iraq from losing the war. To this end, the United States either supplied directly or arranged for others to supply conventional weapons, cluster bombs, anthrax, and other biological weapons materials as well as components for nuclear weapons and equipment to manufacture poison gas.

used banned chemical weapons. In fact, the American government allowed him to buy bacteria and fungus cultures and other items necessary to make biological and chemical weapons. Why the Reagan administration supported Saddam needs to be understood. First, American relations with Iran had soured since Americans were taken hostage at the American embassy in Tehran in November 1979. Pictures of them hooded and bound and the dramatic failure of the rescue mission five months later in April 1980 made anti-Iranian policy popular. But the hostages had been released in 1981. The real reason for American support to Saddam was less emotional: it was the fear that if a vibrant, revolutionary Iran defeated Iraq, it would "destabilize" the whole oil-producing Gulf by inciting fellow Shiis to revolt. Such a "worst-case scenario" would give Iran a virtual monopoly of Middle Eastern oil since even Saudi Arabian oil is produced in the predominantly Shia Eastern Province. It followed, officials of the Reagan administration believed, that American policy should aim at "creating a level playing field" between the Iraqis and Iranians. Since Iran had more people, Iraq needed more arms. America had trained the Iranians; it could supply the Iraqis. As then foreign minister of Iraq Tariq Aziz rightly said to me, "You don't want the war to end. You want to keep Iraq bleeding and not to let either side lose."

What actually happened was that after the Iranians failed in their campaign to take Basra, the Iraqis gradually and painfully regained the advantage. They created new outlets for their oil, pipelines through Turkey and Saudi Arabia, and stepped up exports. They borrowed nearly $100 billion from Saudi Arabia, Kuwait, the Gulf states, and other lenders. The Soviet Union, which had stopped supplying arms to Iraq early in the war, resumed supply on a large scale. Making better use of armor (which was mainly Soviet supplied), airpower (which was partly

French), and intelligence (which was largely American), the Iraqis wore down the Iranians. From 1984 their chemical weapons terrified Iranian soldiers, while their missile strikes demoralized Iranian civilians. As the effects of American attacks on the Iranian fleet in the Gulf and the ban on arms sales to Iran took effect, its revenues dried up and its equipment broke down. Iran faltered. The war was brought to a halt under a UN Security Council–sponsored ceasefire in July 1988. Both sides had lost.

Meanwhile, as Iraqi fortunes in the war against Iran had improved, the Kurdish "problem" simmered. From 1984, the Iraqis had authorized Turkish operations against their Kurdish dissidents, including pursuit of them into Iraqi territory. Then and later, strikes on Kurdish targets within Iraq would be mounted by the Turkish air force. Caught as they are between Iraq and Turkey, the Kurds have been battered by both; which they fear and hate more shifts from year to year. Then it became Iraq's turn. As the war against Iran began to turn in Iraq's favor in the south, Saddam was encouraged to reengage in the north against the Kurds. They had profited from the earlier Iraqi setbacks to renew their struggle and were encouraged and aided by the Iranian government and by the Israelis.

Temporarily, the two Kurdish factions, the supporters of the Barzani clan, known as the Kurdish Democratic Party (KDP) and those of Jalal at-Talabani, known as the Patriotic Union of Kurdistan (PUK), managed to form a more or less united front. By the spring of 1987, they controlled everything outside the major cities. So serious a threat did this appear to the Baath regime that it decided to virtually wipe out Kurdistan. Although previous campaigns had been brutal, the new campaign, known as Anfal, from the name of Quranic verse VIII, wherein the "wicked" are threatened to "Taste the punishment of burning," was to be barbaric. Not only Iraqi troops but Kurdish militiamen, recruited from those with grievances against other Kurds, acted out tribal vendettas and engaged in theft, rape, and murder on a scale not witnessed since the Mongol invasions.

The ensuing chaos was too inviting for the Iranians to resist. Seeing the Kurdish desperation and fury as an opportunity, they advanced into Kurdistan. That spurred the Iraqis to new violence. In March 1988 they counterattacked. A prime target was the town of Halabja, which had recently been taken by Iranian and Kurdish forces; there the Iraqis dropped leaflets warning the population that they intended to use chemical weapons. They did and killed about four thousand men, women, and children. It was a ghastly affair and was rightly condemned, but it was only the most dramatic of many horrible events. The number of casualties will probably never be known, but it certainly ran into scores of thousands, while well over a million people were driven from their homes.

Neither the war in Kurdistan nor the war with Iran solved anything. The one devastated a whole society, but nourished hatreds to be taken up by their followers, while the war against Iran wiped out virtually a generation of Iranians. Taken together, they also halted the promising Iraqi development program that was the bright side of Saddam's rule and encouraged the exodus of more than a million professional men and women on whom Iraq's future depended. None of these disasters seems to have affected Saddam. What worried him was how he could stay in power. That, it is now clear, was always the central theme of his life. To stay in power, he was willing to pay any price, displacing or killing tens of thousands of Kurds and Shiis, purging his party and his army, exiling, imprisoning, or murdering even his closest associates and kinsmen. Staying in power, he decided, depended largely on the amount of money at his disposal: the hundreds of thousands of soldiers made redundant by the ceasefire with Iran wanted jobs; the pampered elite of Saddam's relatives, friends, and supporters wanted public works projects on which they could make money; the entire population was starved for the consumer goods they had come to expect. Rumblings of discontent could be heard everywhere. Had what the great Arab historian Ibn Khaldun termed the "dye of kingship" been bleached from Saddam's banner? Or, as Kipling less poetically put it, had "the wolf missed his kill"? Was Saddam no longer able to govern? That some Iraqis thought so we know because at least two assassination attempts were reported. Perhaps there were many more, because Saddam lashed out not only against the known plotters but also against many others. But he recognized that repression alone would not work; he needed to return to the policy that had worked before the Iran war, "filling the Iraqis' bellies." How to do this was a question of survival. His survival. Getting money was his most pressing objective.

That objective seemed to be receding in 1990. In January Iraq sold its oil for $21 a barrel, but six months later it realized only $11. With the war over, no new loans were forthcoming from other Arab states, and Kuwait was pressing for repayment of funds already spent. Saddam was desperate. His very survival seemed at stake. And right next door was the bank. In these circumstances, he made another miscalculation so colossal that some observers thought he had been set up for defeat.

We do not need to agree with Saddam's view but without taking it into account and understanding the context in which he reached his decisions, we cannot understand what Iraq did. I believe the following is what he thought.

The key to his survival lay in the hands of Kuwaitis. They were acting not as "Arab brothers" but as greedy moneylenders. Disloyal

Iraqis, they had been led astray by British imperialism. Working with their British masters, they had fattened off what was really Iraqi oil. Despite this, Iraq had fought for them, protecting them from the Iranians, and now they not only refused to help but were actively engaged in what amounted to economic warfare against Iraq. By exceeding the quota set by OPEC by nearly 2 million barrels a day, Kuwait and the United Arab Emirates had driven down the price. Kuwait, he charged, had gone even further, stealing oil by slant drilling from the Rumailah field which abutted the frontier, and was trying to prevent Iraq from developing an outlet for its production on the Gulf. This remnant of imperialism must finally be liquidated. After all, even Iraq's kings had demanded Kuwait be "returned." That also had been the policy of Abdul Karim Qasim and the Arif brothers. Had a public opinion poll been taken in Iraq, most Iraqis would have agreed.

Saddam had no way to test his appraisal. He had denied the public any voice in government, and even his closest associates were terrified of voicing any opinion that might run counter to his. He had no agency within his apparatus, like the Policy Planning Council or the National Intelligence Council in the U.S. government. Although an omnivorous reader, he was a man of limited education, with little knowledge of world affairs. Effective government, in his view, did not rest on intelligence and analysis but on manipulation of people and the acquisition of weapons. Such detailed knowledge as he had confirmed this: he had benefited from the fact that the CIA had used Kuwait as its base of operations against Abdul Karim Qasim. Might it not be used similarly against him? He would have to have been naïve not to ponder that possibility, and no one accused him of being naïve.

Weapons, he thought, were precisely what made powers "great" and what gave their governments "security." From the 1970s he had devoted much of Iraq's income to acquiring particularly the "trump cards"—weapons of mass destruction. He had concluded, the record shows, that Iraq would never be treated as a major power or even be secure until it acquired a world-class arsenal.

By nature a secretive man, who had achieved power by conspiracy, Saddam readily accepted what has proven to be a major characteristic of the international balance of power in our times: the existing nuclear powers do not want any new members of their "club." So those seeking to join must go through a period in which they risk attack and almost certainly face opposition. If he was in any doubt about this, he was quickly educated by Israel.

Israel, which already had a full arsenal of nuclear, chemical, and biological weapons and so was a de facto although unannounced member of the "club," carried out a series of spectacular attacks on the Iraqi program. They included the murder of experts Saddam had hired from Egypt and Canada and the destruction of equipment Iraq had purchased from European countries. Then, on June 7, 1981, the Israelis carried out an air strike on the nuclear center just outside Baghdad. While these acts only disrupted parts of the Iraqi weapons program, they convinced Saddam that he must move rapidly and in the utmost possible secrecy.

From a Western perspective, this policy seemed not only dangerous but even immoral. A sober view of recent history shows it was common. America had led the way. In the utmost secrecy and with all possible speed, the Soviet Union followed. Then one after another, France, Israel, China, India, and Pakistan moved to acquire nuclear weapons. Later, North Korea and Iran would undertake their programs. Each government had realized that the period while it was acquiring such weapons was dangerous, but that once it had acquired even a limited stock of weapons, the other members of the "club" would acquiesce

and treat it as a fellow member. So, just as Israel sought the ultimate trump card against the Arabs, and India and Pakistan each
sought it against the other, Saddam felt impelled to seek it. He
apparently thought, probably rightly, that if he actually acquired
nuclear weapons, he would be safe from attack. Many observers
have commented that his major miscalculation on Kuwait was
not the *act* but the *timing*: if he had waited until he had nuclear
weapons, the United States government might have considered
that intervention was too dangerous, as apparently it has concluded with North Korea.*

In any event, Saddam probably thought, after Kuwait was
taken, the other powers would acquiesce as they had done elsewhere. Many people in Iraq pointed out to me that this is what
happened when China invaded Tibet, Indonesia overwhelmed
East Timor, and India conquered Goa. Goa, they said, was a
particularly close parallel. Calling it "properly" a part of India—
which had not existed when Goa was established by Portugal
four centuries before, as Iraq had not existed when Britain established Kuwait—Indian prime minister Jawaharlal Nehru in
December 1961 sent against it some thirty thousand troops. The

---

* The only effective alternative, arms control, was not available. Absent that,
each state that felt itself threatened would do its best to acquire weapons of
mass destruction as a national priority. The most explicit public statement on
acquisition policy was made by the senior Indian nuclear arms policy expert
Jaswant Singh, writing in *Foreign Affairs* in September 1991: "In the absence
of universal disarmament, India could scarcely accept a regime that arbitrarily
divided nuclear haves from have-nots." Knowing that Israel had nuclear,
chemical, and biological weapons and that Iran, with which Iraq was in a state
of war, was on the way to acquiring them, Saddam felt he had to have them. I
imagine that *any* future Iraqi government, even a democratic government,
absent the creation of a regional nuclear-free zone, will try to do the same.

Goans vigorously fought the Indian invaders, but no Western power had objected to Nehru's invasion. As Iraqis bitterly remarked to me, the difference was that Goa had no oil.

A man guided by power politics, Saddam was convinced that morality mattered little to others. What he knew of British and American policies in the past affirmed his view. Both powers had been guided by what they thought were their best interests. From 1982 to 1987, during the Iraq-Iran war, the Reagan administration gave Iraq weapons, money, and intelligence information (illegally hiding what they were doing from Congress in what has been called "Iraqgate") yet they were willing (in the Iran-Contra affair) to provide Iran with missiles to defeat his forces; Britain also was facilitating the sale of arms to both Iraq and Iran. In Kurdistan both Britain and America played both sides and they had tried to assassinate leaders such as Castro, Nasser, Lumumba, and Qaddafi when it suited their purposes. This was a game he understood.

What he did not understand, though there is some evidence that he tried, was the power of public opinion in the West. His brutal policies and particularly his use of poison gas against the Kurds had, by 1990, made him deeply unpopular. That shift in opinion made it possible for a group known as the "Neo-Conservatives" to gain support in the United States. Their motivation was not the preservation of Kuwait but the "security" of Israel. Their reasoning, which was later adopted by successive right-wing Israeli governments under Benjamin Netanyahu and Ariel Sharon, was based on the notion that the Palestinians would never end their struggle for independence (Arabic: *intifada*) until they saw the two major Arab powers, Syria and Iraq, defeated. The more Saddam espoused the cause of the Palestinians, the more dangerous to Israel he became, in their view. By 1990, both the American press and the U.S. government's propaganda radio, the Voice of America, were talking about his overthrow.

So, in that dangerous combination of desperation, ignorance, anger, and greed, Saddam moved toward his invasion of Kuwait. Alarmed at the turn of events, both King Husain of Jordan and President Mubarak of Egypt intervened to mediate and both thought they had averted a crisis. They had not. Saddam sent one of his advisers to the Gulf to see if his pleas for financial help and threats of a change in oil policy had caused a real shift. He came back in the first week of July to report that neither Kuwait nor the United Arab Emirates was likely to adhere to the OPEC quota, and so the price of oil would probably remain low. His report was confirmed a few days later when Kuwait backed out of a deal brokered by Saudi Arabia to adhere to the quota. Iraq then moved troops to the Kuwaiti border.

As was his wont, Saddam had hedged his bets. He knew that his eastern frontier with Iran was secure since the war with Iran had ended and Iran was exhausted; to the north with the Turks he had tacit understandings and effective military cooperation arising from a shared anti-Kurdish policy; and to the west he had moved to repair relations with both the Syrians and the Jordanians. He wasn't then much concerned about Saudi Arabia. What he worried about were the Americans.

On the day he had moved his troops to the Kuwaiti frontier, the State Department spokesperson replied to a question on whether the United States planned to defend Kuwait, "We do not have any defense treaties with Kuwait, and there are no special defense or security commitments to Kuwait." Seeking further assurance, Saddam called in the American ambassador, April Glaspie. What was the attitude of her government to the Iraqi-Kuwaiti problem? he wanted to know. Possibly a clear and forceful statement that America would protect Kuwait would have deterred him. But, under instructions from Washington, which were repeated to other American embassies, the ambassador said in effect that America took no position on frontier disputes among Arab states. This statement was confirmed by the assistant secretary of state in testimony before Congress on July 31, 1990. Taking these statements as a green light, Saddam prepared to act.

He ordered his forces to invade Kuwait on August 2, 1990. Twenty-four hours later, Kuwait was his. But his forces missed their most essential target: the ruling shaikh, who fled the country. Undaunted, Saddam declared Kuwait "returned" to the homeland, as a province of Iraq. Soon he would realize that he had plunged Iraq from the Iranian frying pan into the American fire.

FIVE

AMERICAN IRAQ

The Iraqi invasion of Kuwait inaugurated the period I call "American Iraq" because from 1990 to the present it has been mostly American action that determined events. So fast have these events followed one another that they often appear incoherent or are obscured by what happened next. Here I will seek to explain their inner relationships and to show how a pattern has emerged.

The initial American response to the Iraqi invasion of Kuwait was indecisive. This was the result of two themes that would run through the fifteen years of "American Iraq": misunderstanding and deception.

As we have seen, the "signals" sent to the Iraqi leader during 1990 constituted what he took to be American acquiescence in his attack on Kuwait. They at least began with misunderstanding. While the relevant documents are still unavailable and in any case may not reflect the thought process of the American officials, I think that they will ultimately show that American officials thought that Saddam just wanted to settle the long-standing dispute over the Iraq-Kuwait frontier. Frontiers mattered, particularly in Iraq's far south where it had only a small door to the Gulf; Iraq had already gone to war with Iran partly over that outlet.

The Kuwait frontier dispute figured in a discussion I had with the then foreign minister of Iraq, Tariq Aziz, in 1983. He and other Iraqi officials had argued that the way the frontier had been drawn hampered Iraqi access to the Gulf, and they raised it periodically with the Kuwaitis. Iraq needed access, and what it wanted was of little value to the Kuwaitis. To push the Kuwaitis to negotiate is the best explanation for repeated statements that the U.S. government was not concerned with frontier disagreements among Arab states. If the Kuwaitis did not yield, it may be that the first Bush administration was willing to tolerate a limited Iraqi seizure of this desolate strip. This interpretation was substantiated in an interview Ambassador April Glaspie gave to the *New York Times* just after the invasion. She said no one in the American government thought the Iraqis were going to seize *all* of Kuwait.

If American officials were surprised by the seizure of all of Kuwait, they were poor students of both history and economics. For eighty years the Iraqis had regarded Kuwait as a remnant of imperialism, "artificially" split off from what became Iraq and so legitimately a part of Iraq. Kuwaitis had begun to disagree only after they had become rich on oil. When I lived in Baghdad in the 1950s, before the major impact of oil, Kuwait was then just a little trading and fishing village with no paved streets, no electricity, and no running water. Its people looked to Iraq for all they needed, their luxuries, their schooling, their pleasure. I once entertained for dinner its shaikh, who had come to Baghdad to buy the shotgun shells he could not find in Kuwait.

Then came the flood of oil money. It brought to the fore the

Kuwaiti merchant tradition, the same tradition that Muhammad had long before rebelled against in Mecca. Money was Kuwait's god just as it had been Mecca's. The little city-state of Kuwait used its money to win friends among its Arab "brothers" just as Mecca had bought off the bedouin. Kuwait did it on a grand scale. One of the real statesmen of the Middle East, Abdul Latif Al Hamad, created a clone of the World Bank, the Kuwait Fund for Arab Economic Development, that pumped billions of dollars into Asian and African states. Other billions were given out directly by the Kuwait government. Iraq was one of the beneficiaries. But, as a consequence of the war with Iran, Iraq's need was insatiable; it required much more than Kuwait was willing to give. When the war ended and the danger of an Iranian invasion dissipated, Kuwait no longer needed Iraq. As bankers, Kuwaitis cast a shrewd eye on Iraq and found it a poor risk. So they refused further loans and demanded repayment of what they had lent. Trying to open their purse, Saddam first appealed to Arab brotherhood, next he pleaded, then he threatened, and finally he invaded. He had to take Kuwait because, as a famous American robber once replied when asked why he robbed banks, "That's where the money is."

Saddam had an old-fashioned view of how money is kept. He hoped that he would find vaults full of vast treasures. That is how money used to be kept when Hulagu Khan sacked Baghdad in 1258, the British plundered Bengal in 1758, and the French looted Algiers in 1831. Unfortunately for Saddam, the banking system had changed. Gold and jewels were no longer piled up in the treasury. He did find about $2 billion in the central bank, but the bulk of Kuwait's assets—hundreds of billions of dollars—was beyond his reach in the international financial system. There it was quickly frozen by the United States and other powers.

In the United Nations and elsewhere, the United States and Britain, occasionally supported by the French and Russians, condemned the Iraqi aggression. The Iraqis were not impressed. A believer in realpolitik, Saddam assumed that they would make pro forma protests but soon accept "facts." That is what they had done recently in India, Indonesia, and Tibet. He also scoffed, with reason, at the charge of immorality; his evidence was impressive: He knew all about "Iraqgate" because he was the beneficiary of American money, arms, intelligence information, and diplomatic support. He even got materials and equipment to make chemical, biological, and nuclear weapons from America and Britain[*] while they publicly proclaimed the immorality of having or using them. He also knew about "Irangate," when the Americans gave Iran weapons to use against him. Like everyone in the Middle East, he knew that the CIA and MI-6 had overthrown the democratically elected government of Prime Minister Mossadeq in Iran. As an insider, he knew that the CIA at least helped to overthrow the regime of Abdul Karim Qasim and was involved in the bloodbath that followed. He knew that the CIA, MI-6, and Mossad promoted assassinations of statesmen and helped to "destabilize" their governments. In the "mean street" of world politics, the Western powers had shown scant regard for

---

[*] Largely ignored by the Western press, a list of the companies involved, as reported by the German newspaper *Die Tageszeitung* on December 19, 2002, included Honeywell, Unisys, Sperry, Rockwell, Hewlett-Packard, DuPont, Eastman Kodak, Bechtel, and many European companies. It can be accessed at www.taz.de/pt/2002/12/19/a0012.nf/text. The U.S. Senate Banking Committee turned up dozens of biological agents shipped to Iraq under license from the Commerce Department, including various strains of anthrax. See the *Washington Post*, December 30, 2002. The U.S. government also supplied cluster bombs through a Chilean front company.

morality. So, although Saddam was surprised by the immediacy, vigor, and uniformity of the reaction, he continued for months to believe that the reaction would remain only talk.

What apparently Saddam did not properly judge was that he had put his hand on two things where the Great Powers would not tolerate interference: money and oil. They, not morality or legality, were what differentiated Kuwait from Goa, Tibet, and East Timor. As an American congressman then wryly remarked, "if Kuwait produced bananas instead of oil," Saddam's grab might have been tolerated.

The conquest itself had met little resistance; Kuwait could not match the large, battle-trained, and heavily armed Iraqi army. Kuwait was only a city-state on the scale of ancient Babylon or Athens. But, quick as they were, the Iraqis failed to catch the ruler—he flew in an American military helicopter to Saudi Arabia, where he was able to establish a rudimentary government-in-exile around which it was possible to assert a legal challenge to the Iraqi occupation. Meanwhile, Iraq fumbled. First it set up a quisling government under a junior army officer related to the ruling family, but that government lasted hardly long enough to acquire a name. Then Saddam decided to peel off the long-disputed northern strip, which he incorporated into the province of Basra, and proclaimed the remainder Iraq's nineteenth province.

With unprecedented speed, Western surprise had worn off. At the United Nations Security Council (UNSC), within hours after Iraqi troops moved, the United States secured, by a vote of 14–0, the passage of Resolution 660, demanding immediate Iraqi withdrawal. Even more impressive, the old adversaries, the United States and the Soviet Union, issued a joint statement denouncing the Iraqi invasion. Saddam had probably expected and discounted these moves. More serious, however, was the resolution the Security Council passed a few days later, 661, calling for a boycott of Iraq's overseas trade.

In response, Saddam put forward his first proposal to end the crisis. In the months that followed, he put forward others. Jordan, Morocco, Libya, France, and Yugoslavia each came up with proposals. All were rejected by the United States and/or by Iraq. Perhaps the most serious attempt to stave off war was that of a member of Mikhail Gorbachev's "Security Council" (and later prime minister), Dr. Evgeni Primakov. During October, Primakov made two visits to Baghdad. As he later told me, he got Saddam to agree to withdraw from Kuwait on two conditions: first, that American forces also withdraw, and second, that an international conference be assembled to resolve all the outstanding problems of the Middle East, including nuclear arms and the Israeli-Palestinian conflict. Primakov then flew to Washington where on October 19 he briefed President George Bush on Saddam's offer. Bush, Primakov told me, looked surprised and said, "This is the first I have heard of such an offer. How long will you be in Washington?" Primakov replied, "As long as necessary, Mr. President." Bush then said, "Give me some time and I will get back to you." Primakov was having lunch at the presidential guest house when one of the president's aides came in and said, "You might as well pack your bags."

In fact, the Bush administration had already determined on war and was unwilling to negotiate an end to the crisis. It regarded the offer Primakov conveyed as a "derailment" of its policy.

Neither in the United Nations Security Council nor elsewhere did Iraq get useful diplomatic support. While the Saudis partly blamed Kuwait for the crisis, they were disturbed because the Iraqis had replaced a "royal" regime with a "republican" regime—for Saudi royalists, republicanism was sedition. Monarchy, republic, or dictatorship, most other Arab governments condemned the Iraqi invasion; in an Arab League meeting in Cairo, a majority voted against Iraq. More serious for Saddam was that both the Saudis and the Turks closed the pipelines passing through their territories so that Iraq's oil outflow—and money inflow—were severely diminished.

Meanwhile in Kuwait, Iraq followed the brutal pattern it had adopted in suppressing its Shiis and Kurds—destruction of property, arrests, torture, and executions. Perhaps as many as one thousand Kuwaitis were executed as enemies of the (new) state. To these terrible crimes, it added widespread looting. What it was doing was documented and publicized by Amnesty International, thus creating a view of the Baath regime as Nazi-like. As though it had picked the most maladroit moves it could find, the Iraqis forced tens of thousands of Asian workers to return home and took as hostages the male members of the expatriate Western community, planning to use them as "human shields" in the event of Western attack. Compounding its "image" problem, it allowed these hostages to leave, as requested by former British prime minister Edward Heath, British member of Parliament Tony Benn, and former German chancellor Willy Brandt, in small groups. This constant activity kept Western media focused until December 1990, when the last group was sent home.

During these months, the United States government was urging the Saudi Arabians to "request" the stationing of American troops in their country. King Fahd was reluctant to do so both because Arabia, the "cradle" of Islam, had long considered itself an Islamic preserve and because his very conservative government believed that the population would be offended by Western customs. After a great deal of diplomatic arm twisting, the king agreed to invite Defense Secretary Dick Cheney to

Riyadh. Then events took on a life of their own; by the first days of 1991 nearly 250,000 troops, at least 1,000 aircraft and about 30 naval ships capable of firing missiles or launching aircraft had been assembled in the Gulf area. In addition, long-range B-1, B-2, and B-52 bombers were stationed within range of Iraqi targets.

The Americans also began the laborious process of rounding up allies. To that Bush administration, allies seemed important. They paid much of the cost of the war (on which America actually made a small profit), and the presence of Arab troops, even in token size, helped Fahd acquiesce in the stationing of foreign troops in Arabia. Egypt was important because, just as the British had found in the First World War, the Suez Canal was the safest and fastest route to Arabia. To win over Egypt, which was as strapped for cash as Iraq, the Bush administration did for President Mubarak precisely what Saddam Husain had asked Kuwait to do, forgive its debts and provide more money. In various forms, the total amounted to many billions of dollars. Turkey, where the air base of Incerlik was situated, got a huge consignment of military equipment, loans, and preferential trade arrangements. Syria, the other "bad boy" of the Arab world, was given money, arms, and a license to continue its intervention in Lebanon. Even the Soviet Union was helped to get several billion dollars' worth of loans, credits, and cash from Saudi Arabia and the Gulf oil states. The main loser was Yemen, which had opposed American policy: America stopped its aid program, while Saudi Arabia expelled nearly a million Yemeni workers on whose remittances it heavily depended. With the coalition shaping up, the Bush administration on November 29 secured UNSC Resolution 678, setting January 15, 1991, as the deadline beyond which "all necessary means" would be used to effect Iraqi withdrawal.

Despite last minute diplomacy, including a tense, six-hour

meeting between U.S. secretary of state James Baker and Iraqi foreign minister Tariq Aziz in Geneva, Saddam refused to give in. Possibly he still believed that the coalition would fall apart or that the Americans would back off; probably he feared that if he was seen publicly to turn tail and run, he would be overthrown by his own army; certainly he miscalculated. At the eleventh hour, January 13, 1991, UN Secretary General Javier Pérez de Cuéllar flew to Baghdad, but Saddam would not budge.

War came on January 17, 1991; like the 2003 American invasion, it was never really a contest. Iraq then had a larger army than in 2003, but it was armed with obsolescent military equipment, was weak in command-and-control, and was almost totally without high-tech weapons. The Iraqis had nothing to match the huge aerial armada that would fly more than 106,000 sorties and drop 88,000 tons of bombs. Nearly 300 Tomahawk guided missiles, each carrying half a ton of high explosives, were also fired at Iraqi targets. This massive air assault pulverized Iraq before any ground troops engaged.

In impotent fury, the Iraqis fired a few missiles at Israel, presumably calculating that, if the Israelis retaliated, the Arab coalition members would withdraw. They also aimed a few missiles at Saudi Arabia. None did significant damage. The truly horrible damage was done in Kuwait where beginning on January 22, some seven hundred oil wells were set afire and oil was allowed to pour into the Gulf where it created a 350-square-mile, or 900-square-kilometer, slick.[*]

---

[*] Subsequently, Evgeni Primakov and I arranged a joint Russian-Western venture to put out fires in the Kuwait oil fields.

Even at this point, attempts were made to bring the fighting to a halt. Massive popular demonstrations were held throughout Europe and America, Pope John Paul II condemned the war, and President Gorbachev brought forward another peace plan. President Bush was, apparently, very disturbed by events or proposals that might deflect the assault. In the account he and Brent Scowcroft wrote on his presidency, *A World Transformed*, Bush recounted that "Early in the morning of Friday, February 15 [1991], one of the White House staff came to our bedroom, where Barbara and I were reading the papers and drinking coffee, and reported he had heard there would be an announcement at 6:30 from Iraq. I turned on the TV and we anxiously waited as 6:30 came and went. Finally, an anchor cut in and reported that the Iraqis had announced they would comply with Resolution 660, including the withdrawal from Kuwait. Instead of feeling exhilarated, my heart sank." Bush wanted to go to war.

On February 24, judging that the Iraqi army had been effectively suppressed, the Americans began the ground offensive. Already on February 25, the Iraqis began to withdraw, but they were slaughtered on the "road of death." Nothing on that scale of massacre had occurred in Middle East wars since Hulagu Khan took Baghdad. Saddam tried to negotiate terms but finally capitulated on February 27. President Bush then ordered a ceasefire. The toll was immense: perhaps ten thousand civilians and thirty thousand Iraqi soldiers were killed. Proportional to the population, that was more than five times the casualties suffered by America in the Vietnam war. But some units of Saddam's elite Republican Guard managed to escape. Ironically, both the Iraqis and the Americans had "missed their kill": had the shaikh of Kuwait not eluded the Iraqis, restoring his regime would have been far more difficult, and had the Republican Guard not eluded the Americans, Saddam might not have survived.

President Bush was roundly criticized for stopping the American army short of Baghdad, but, as he wrote in his account of the events, *A World Transformed*, "Had we gone the invasion route, the United States could conceivably still be an occupying power in a bitterly hostile land." As it turned out, his comment was not only a justification but a prediction.

On March 2, UNSC Resolution 686 required Iraq to pay reparations, release all prisoners, return all looted property, and void all edicts it had promulgated on Kuwait. Iraq could not return the stolen property that had been passed out to its population, and it was unable to pay reparations unless its shattered economy was at least partially restored. But the Iraqis could not equivocate. They accepted.

Meanwhile, two revolts had broken out against the regime in Iraq. Beginning in Basra the first of March, rebels, primarily members of the Iraqi Shia community, took part in an antigovernment "uprising" (Arabic: *intifada*). Apparently it was spontaneous and was sparked by deserting Iraqi soldiers. It was encouraged by both the Iranian and American governments. However, neither gave the rebels effective aid. On the contrary, the American commander, General Norman Schwarzkopf, allowed Saddam's regime to use helicopter gunships against the rebels. On the ground, the American forces allowed attacking Iraqi army units to pass unopposed through their positions and even defended arsenals to prevent Shiis from arming themselves. Remembering American hostility to the Iranian Shia government, Iraqi Shiis believed Schwarzkopf's move was a deliberate move to weaken them.

The dissident Shiis were weak. Their revolt lacked not only arms and mobility but, more important, coordination: it was a neighborhood rather than a national war in which many devoted themselves to destroying government property, looting, and carrying out vendettas. The only overall leadership came from the Shia religious establishment (Arabic: *al-marjiiya*), which for years had struggled against the regime. Although decimated by Saddam's security forces, it retained the respect and loyalty of the Shia community, but the Shiis could not stand against the highly trained and well-armed Republican Guard. In town after town, the guard beat them down in scenes of truly horrible repression. By March 25, 1991, they were overwhelmed.

Outside the country, several attempts were made to set up broad opposition fronts. But, after years of discussion about remaking Iraq, the participants demonstrated that they could not work together. Many were completely out of touch with trends within the country and even with the people they aspired to lead. They were also, by and large, out of sympathy with the Shia clerics who were the de facto leaders of the revolt. These conditions were to be repeated in 2003.

It was about this time that Saddam put into motion a plan that had been discussed for decades under various Iraqi governments, the draining of the vast marshlands at the southern end of the country. In Abbasid times, the marshes had been Iraq's "Siberia" where the Zenj, blacks imported from Zanzibar, slaved away mining minerals. Over the centuries the "Marsh Arabs" developed a distinctive culture, a mode of life governed by their habitat and by an affinity to Shiism. During the war with Iran and the more recent civil war, large numbers of army deserters and opponents of Saddam's regime had taken refuge there. So, although there were economic reasons for draining the marsh, Saddam's motives were to destroy the refuge, stifle the activities of the Shia inhabitants and prevent them from hiding army deserters. Today, the 8,000-square-mile, or 20,000-square-kilometer, area is a wasteland, drained both of its marshes and its people.

Without any coordination with the Shiis, the Kurds also mounted yet another rebellion. Having been fighting sporadically for decades and benefiting from access both to sanctuaries within Iraq and to safe havens in Iran, they were better armed and more organized. Their strength increased when the "tame" Kurdish militia, known as the "the horsemen" (Arabic: *fursan*), who had been armed by the regime to fight the "wild" Kurds, joined the rebellion. The Kurds also benefited from having both a tactical and a strategic aim: the tactical aim was to capture the city that housed the oil industry, Kirkuk (which they did on March 20), while their strategic objective (in which they failed) was to win independence.

Like the Shiis in the south, the Kurds were astonished by the speed and force with which the Iraqi army was able to deploy. Before the end of March, it had retaken all the major centers. When no outside aid was forthcoming, the Kurdish *intifada* collapsed. Then about one in two Kurds fled. Many tried to cross into Turkey, but the Turks closed their frontier; they did not wish to add to their own Kurdish dissidents. Finally, in April, at the urging of British prime minister John Major, President Bush authorized a relief mission known as "Operation Provide Comfort." The Kurds did not get much comfort.

Unlike the Shiis, the Kurds had neighbors who were accustomed to meddling in their affairs. Iran alternatively aided them, sold them out, and fought them. Turkey went further: it denied that any such people as Kurds existed—Kurds were referred to as "mountain Turks." The Turkish government attempted to suppress Kurdish language and culture and, more violently, engaged in an often brutal military campaign, employing American-supplied weapons, against them. Turkey's policy differed from Iraq's only in that it was somewhat less genocidal.

The Kurds had no illusions about the Iranians or the Turks but were sorely disappointed when the United States first encouraged them and then let them down, as it had done in the 1980s. Given the reality of their position, stretched over Syria, Turkey, Iraq, and Iran, the Kurds knew they had little hope of achieving independence. But they also believed that without at least autonomy, they would become an endangered species. This was a correct assessment. It posed in 1991 and poses in 2005 the parameters of their dilemma. Revolt against Iraq did not succeed; independence would not be allowed by others; so the Kurds wavered restlessly between fighting and fleeing, opposing and cooperating, working together and feuding. Kirkuk remained, as it had been during "British Iraq," the prize: it had the oil field on which all hopes depended. When the Iraqis secured it, they withdrew to a defensive line on the plain, while the Kurds stayed in their mountains. For a decade neither side supported the arrangement, but both acted as though it was permanent. The Kurds got de facto independence, and the Iraqis got time to rebuild their forces.

On April 3, 1991, Security Council Resolution 687 set out what has come to be known as the sanctions regime. Under it Iraq was charged with dismantling weapons of mass destruction and facilities to manufacture them. The Iraqi government was to allow the United Nations Special Commission (UNSCOM) to monitor its compliance on chemical and biological weapons and act in conjunction with the International Atomic Energy Administration (IAEA) on nuclear weapons. Since both the requirements and the authority to enforce them represented major intrusions into Iraqi sovereignty, the Iraqi government periodically tried to delay, thwart, or disobey them; when it did, the United States and Great Britain (with occasional participation by France) reacted with airstrikes or threats to invade.

Although foreign ground troops were restricted to frontier areas, the potential for wide-ranging air intervention remained. There was no explicit UN authorization for control of Iraqi airspace, but a northern "no-fly" zone was established in April 1991, ostensibly to protect aircraft delivering relief supplies to the Kurds. Prohibition of Iraqi flying north of the 36th parallel was to last, under various names, until 1998. In the south, a similar zone was created four months later. It specified that no Iraqi aircraft were to fly south of the 32nd parallel. Thus, all but a band across the middle of Iraq was declared off-limits to Iraqi aircraft.

Apart from weapons and restrictions on air activity, the Security Council imposed a boycott on Iraq and created a committee to oversee its application. Under its terms, all Iraq's financial assets abroad were frozen, and both imports and exports were banned, except for medical supplies and certain food products under the "Food for Oil" program, begun in 1995. The original resolution was subsequently expanded to restrict both sea and air transport and to authorize the committee to determine when or if food was to be imported. Critics have termed it one of the most draconian and punitive measures ever imposed on a defeated power. It would remain in effect for seven years. It severely harmed the Iraqi people but did not prevent the government from buying arms.

While the resolutions had not specified the objective of bringing about "regime change," it was inherent, since as long as the sanctions were in force, the regime could not meet the specified requirement to pay reparations. Like a catch-22, the effect was circular: no lifting of sanctions until reparations were repaid and no possible repayment until sanctions were lifted. Regime change was intended to be the only way to break the circle. Regime change—that is, the overthrow or killing of Saddam Husain—was openly proclaimed as the American objective by President Bush.

Experience in Iraq during the 1990s showed the weaknesses of sanctions against a strong and determined government. The Iraqi regime was able to deflect their impact so that the general population rather than its core supporters suffered; the population indeed came to hate those who had imposed the sanctions rather than those whose actions had occasioned them. Since the sanctions aimed at destroying the regime, Iraq reacted naturally with continued efforts to arm itself. And what the sanctions did to Iraq also spilled over into neighboring countries. Jordan, whose largest trading partner was Iraq, suffered significant losses, while Turkey, through which a major oil pipeline passed, estimated that it lost nearly $30 billion when it was forced to close the line. (Both countries were tacitly allowed to violate the sanctions regime.)

Fearing his army and increasingly unsure even of the party structure he had so laboriously created, Saddam Husain turned back to a policy first set forth by the British in the 1920s and elaborated by the parliaments they created. The essence of this policy was reliance on the leaders of clans, the people the British had "promoted" to be "chiefs" and who, in conjunction with city moneylenders and merchants, had driven their fellow tribesmen down into serfdom. As resurrected by Saddam, the policy involved passing out land deeds and money to the reconstituted chiefs. The move was symbolized by reinstating the use of tribal designations (Arabic: *laqabs*) in personal names. This was a practice that had been banned, as a residue of "feudalism," a generation before. Once again, as in the 1930s, tribal "chiefs" took over the national assembly. But, as with everything else he did, Saddam twisted the system to his advantage. He gave the chiefs he appointed money and arms and put them to work to watch even the Baath Party; thus he created a new form of tribalism on top of the old. At the center of this new organization of prestige and power was the clan of Saddam himself, al-Majid, which was a part of a larger, more diffuse, less closely related tribe known as the Al Bu Nasir.

What he was doing was falling back on the form of social organization of pre-Islamic Arabia (Arabic: *al-jahaliyah*). What he was proclaiming—and demanding—was the absolute loyalty effected in a clan (Arabic: *qawm*). In this tight kinship group, all members were jointly responsible for the actions of any single member and each was required to take vengeance on anyone harming any other member. While it is doubtful that Saddam knew enough history to place his new policy in context, he at least unconsciously did. Like all Iraqi children, he had learned his language by memorizing poems dating from the *jahaliyah* in which the imperatives of loyalty, honor, and hardihood are focused on the clan. He would have had to be blind and deaf not to have absorbed this message even in the rudimentary education he had received.

So it was that when driven to the wall by his defeat in the Kuwait war, shaken by the massive revolts among the Shiis and Kurds, and coming to distrust even his Baath associates, he fell back on this most primitive of Arab political concepts. In the latest shake-up of the government, all the key posts in the military, the police, and the various competing security organs were assigned to members of Saddam's own clan, the al-Majid. As more distant relatives and nonrelatives were pushed aside, many were purged, imprisoned, or executed so that Saddam, apparently, felt that at last he had arrived at the unshakable core of his power.

It came as a profound shock to him when, in 1995, the al-Majid husband of his eldest daughter packed up his wife, their children, and his wife's sister and her husband and departed for Jordan. This sudden break probably resulted from a clash between him and Saddam's wild, half-crazy, and violent son Udai. As Saddam should have known, such departures are the traditional way that conflicts within a clan were resolved: the weaker party departs and throws itself on the hospitality of another group. That is what General Husain Kamil, his brother, and their wives did. They left and asked for protection from the king of Jordan. Foolishly, the émigrés voluntarily returned to Baghdad, where, apparently, they thought they would be forgiven. Saddam did not forgive them. He never forgave. He arranged that other family members execute them for having brought shame on the clan. Having done that, Saddam also, too late, moved to restrain his son Udai and to replace him with his younger, shrewder brother Qusai.

While the issue of Iraq's weapons program was not a major feature of the shattering of Saddam's family, Husain Kamil's defection necessarily highlighted that program. He was quoted as admitting that Saddam was "cheating." Of course he was, since he knew that the United States was trying to kill him and wanted to overthrow his government. He had always believed that having weapons of mass destruction was his major hope, first to defeat Iran, then to prevent others from taking Kuwait back from him, and finally to prevent his regime from collapsing. While most accounts focus on his deception and obstruction, he was smart enough to realize that *having weapons* was very different from *trying to get them*. He would certainly have liked to have them, but he had realized by the early 1990s that trying to get them was simply too dangerous, and so he gave up his program. During his time in Jordan, where he was free to speak, General Kamil was quoted as denouncing Saddam's weapons program, for which he had been responsible, but what he actually said was quite the contrary. He said that Saddam had actually destroyed such weapons of mass destruction and the means to make them. Parallel to the sanctions program, which the United States, at least, sternly enforced under both the Bush and Clinton administrations, the United States engaged in attempts to destabilize the Iraqi regime. Through congressional appropriations under the Iraq Liberation Act, the United States openly subsidized a number of exile groups that aimed to overthrow the regime.[*]

There were many charges brought against Saddam during

---

[*] The main beneficiary was the Iraqi National Congress, headed by the Ahmad Chalabi. The other main Arab group, long an "asset" of the CIA, was Iraqi National Accord, headed by Iyad al-Allawi, whom the American government later installed as interim prime minister. More significant were various Kurdish groups—the Islamic Movement of Iraqi Kurdistan, the Movement for Constitutional Monarchy, the Kurdistan Democratic Party, and the Patriotic Union of Kurdistan—but these groups disagreed so violently with one another that one was prepared to work with Saddam. Ineffective against him, they convinced Saddam that he would be foolish to work honestly with the United States toward "peace" when his opponents defined peace as being achieved by his death.

the years after the 1991 Gulf War, contacts with terrorists, involvement in the attack on the World Trade Center, masterminding the 9/11 terrorist attack in New York, attempting to acquire nuclear weapons by the purchase of centrifuge tubes and "yellow cake" (uranium oxide), all of which proved to be untrue. Most damning of all was the allegation that the Iraqi intelligence services had tried to murder former President George Bush during a visit he made to Kuwait in April 1993. That allegation was based on highly dubious information but was used by men pushing their own agendas to justify them.*

Saddam certainly engaged in assassination both inside Iraq and abroad when doing so was to his advantage. As he told the American ambassador on the eve of the Kuwait invasion, "We cannot come all the way to you in the United States, but individual Arabs may reach you." But the plot against Bush is inherently unlikely and was based on testimony extracted under duress from one Iraqi common criminal, a smuggler. There is no supporting evidence, and the story the smuggler was forced to affirm was comical. However evil Saddam was, he was not silly. At that time, he had nothing to gain and everything to lose by assassinating the former president because he was then engaged in sensitive negotiations over exporting oil. Kuwait, on the contrary, had something to gain by "discovering" a plot. It wished to cause the negotiations over Iraqi oil to break down.

---

* During the Clinton administration, the two officials were members of the National Security Council staff, Martin Indyk and Samuel Berger; later, during the second Bush administraion, a much larger effort was organized by members of the Neo-Conservative clique who were led by Paul Wolfowitz and were mainly in the Department of Defense and the office of Vice President Dick Cheney.

The Kuwaitis had a history of fabricating events* when they wished to have the United States punish Iraq. They succeeded. With this highly dubious story as its justification, the Clinton administration ordered the firing of twenty-three Tomahawk guided missiles on the headquarters of the Iraqi intelligence agency in downtown Baghdad.

Three things can be inferred from this episode: first, American officials arguing for specific policies could achieve their objectives more easily by tying them to dramatic (even if dubious or untrue) events than by logical argument; second, a president could gain public approval—President Bill Clinton's approval rate jumped eleven points when he ordered the missile attack—by being "tough"; and, third, Saddam Husain, who obviously would have known the truth of the story, would see it as proof that, regardless of what he did or did not do, the American government was intent on destroying him.

It was. The CIA organized flights of unmanned aircraft over Baghdad to drop leaflets urging revolt and spent millions of dollars encouraging plots and coups. The director of one CIA program told the *Washington Post* that it worked with the Iraqi National Accord (Arabic: *al-Wifaq al-Watani al-Iraqi*) to engage in a series of car bomb attacks and assassinations. It failed because

---

* The then minister of information, Shaikh Saud Nasir as-Sabah, who briefed the press on the plot, had masterminded a similar episode during the Iraqi invasion of Kuwait. As ambassador to Washington, he had his daughter testify before Congress that Iraqi soldiers had ripped newborn children out of incubators and flung them on the floor of a hospital to die. That story was false. Helping the ambassador and his daughter was Victoria Clarke, then head of an advertising agency, who later became the Pentagon spokeswoman in the second Bush administration.

al-Wifaq was "penetrated" by Iraqi intelligence, which "rolled up" and executed most of its members in June 1996. (The then head of al-Wifaq, Iyad al-Allawi, himself a former senior Baathist official and intelligence agent who had turned against Saddam, was not caught, and later, in June 2004, became Iraq's first American-appointed interim prime minister.)

In August 1998 even the United Nations Special Commission was compromised when Saddam learned that it was being used to disguise activities by CIA, MI-6, and Mossad agents. UNSCOM was forced to leave Iraq. Infuriated but not denying the Iraqi charge, the United States and Britain retaliated with a major bombing campaign ("Operation Desert Fox") that was opposed by most Arabs, France, Russia, and other states. Saddam could not have missed the significance of these events: a vigorous and wide-ranging clandestine war was backed up by airstrikes. Soon it would become clear that these were precursor moves toward full-scale war.

Iraq hit bottom in 1994. Restricted by sanctions, living conditions had become desperate. Hospitals had run out of medicines and even soap to wash bedding; malnutrition was widespread; infant mortality rates soared; even clean drinking water was difficult to get and was unavailable in many areas. Runaway inflation virtually wiped out the new middle class. Even in those hard times, however, the regime spent heavily on rearmament and pampering those on whose loyalty Saddam depended. Then, two years later, despite the sanctions, the clandestine attempts to overthrow the regime, and serious air and missile attacks, Iraq began to recover. Bridges again crossed the rivers; trade increased; electrical power stations and transmission lines were restored; water was distributed; and sewage was being treated. Iraq began to export oil in December 1996, and by 1998 it was being allowed under the UN Food for Oil program to export nearly $10.5 billion worth. Agreement was reached in that year with Syria to reopen the pipeline to its port on the Mediterranean. The next year, the UN gave up all control over oil exports. By the year 2000, Iraq was earning more than $30 billion.

Important as this economic change was, perhaps as important to Saddam's regime was the collapse of the Kurdish opposition. What happened was another sequel to the tragic history of that fierce and independent mountain people: long divided by their high mountains into separate valleys, the Kurds split into tribal, linguistic, religious, and ideological factions. Sometimes they briefly worked together against common enemies, but usually they so hated one another that each sought outsiders to give it power to destroy the others. For several years the Iranians had been trying to suppress their Kurds. In the usual way, they sought to use one group of Kurds against others. As proxies, they picked Jalal at-Talabani's Patriotic Union of Kurdistan (PUK). In the summer of 1995 Talabani allowed a large force of "Revolutionary Guards" (Persian: *Pasdaran-i Inqilab*) to enter Iraq through territory he controlled. The Iranians did not stay long, but their move encouraged the PUK to team up with a Turkish Kurdish movement, the Kurdistan Workers Party (Turkish: *Partiya Karkari Kurdistan*) to attack Talabani's Iraqi Kurdish rival, the Kurdistan Democratic Party (KDP). Fearing that his enclave of "Free Kurdistan" was about to be overrun, the KDP leader, Masud Barzani, turned to the only ally he could find, Saddam Husain. Adept at the same game Iran played, Saddam was delighted to help the Kurds kill one another. During the first week of September 1996, working with the KDP, Iraqi troops and intelligence forces retook the areas the

PUK had seized and, more important to the Iraqis, arrested or killed many of Iyad al-Allawi's National Accord, which was working with the CIA. Content with what had been achieved, the Iraqi forces then withdrew.

The Clinton administration felt impelled to retaliate, but finding no target in the north, it hit the uninvolved south with forty-four cruise missiles and extended the "no-fly" zone one parallel north. So inappropriate was American retaliation that it was strongly criticized even by Saudi Arabia. Saddam regarded the episode as a victory; he thought he could see "light at the end of the tunnel." The Kurds, as usual, were at one another's throats, and the Americans did not seem to know what to do. The war against him seemed to have sputtered out.

Saddam's assessment of victory was shared by a new group that came to power in Washington in 2001. Occupying what Lenin would have called Washington's "heights of power," key positions in the Department of Defense, the office of the most active vice president in American history, the media, and an assortment of well-funded think tanks, the "Neo-Conservatives"* soon began to implement the American "crusade" they had advocated over the previous decade. For them the campaign in Afghanistan against the Taliban supporters of Usama bin Ladin's al-Qaida was a sideshow. Iraq was what really mattered. It was to be the first step in a perpetual war against any group or nation-state that could contest American supremacy. Supported

---

* I have written six articles analyzing them, their background, their influence, and their aims. They are tangential to my purpose here but can be accessed on my Web site, www.williampolk.com.

by President George W. Bush, Vice President Dick Cheney, and Secretary of Defense Donald Rumsfeld, they launched a vigorous campaign to convince the American public that the Iraqi regime was armed with nuclear and other weapons of mass destruction, had the means to deliver them, and posed a mortal threat to America. Following the attack on the World Trade Center and the Pentagon, they also encouraged the belief that, in league with al-Qaida, the Iraqis were behind it.* Copying the emotional process (based on the alleged assassination attempt on President Bush) that had convinced President Clinton to attack Iraq with guided missiles in 1993, they used the emotional reaction to the al-Qaida attack to implement their program. That program was to lead the United States to war in Iraq. Effecting a "regime change" in Iraq was secretly adopted as American policy the day after the al-Qaida attack on September 11, 2001, and was openly espoused by 2002.

As the slide toward war became evident, the Iraqi government decided to cooperate with a new inspection. Under a Security Council mandate, the United Nations Monitoring, Verification and Inspection Commission (UNMOVIC) was created under Hans Blix, the Swedish diplomat and former head of the International Atomic Energy Administration (IAEA). His team began work in November 2002. Iraq proclaimed that it had no weapons of mass destruction, and UNMOVIC found

---

* All of these assertions have been proven wrong. The definitive study of the weapons issue is Joseph Cirincione, Jessica T. Mathews, and George Perkovich, *WMD in Iraq: Evidence and Implications* (Washington, D.C.: Carnegie Endowment for International Peace, 2004). An excellent documentary is Robert Greenwald, *Uncovered: The Whole Truth about the Iraq War,* www.truthuncovered.com.

none. But the Bush administration repeatedly pressured American intelligence agencies to certify that Iraq had these weapons. When the State Department's Bureau of Intelligence and Research, the CIA, and the Defense Intelligence Agency (DIA) could not say what the Neo-Conservatives wanted to hear, they established their own intelligence agency, the "Office of Special Plans" under a Neo-Conservative director, to proclaim it. After the invasion a twelve-hundred-strong team of inspectors and the Seventy-fifth Exploitation Task Force from the Pentagon found no evidence of such weapons or delivery systems. Their reports were confirmed by the Bush administration-appointed arms inspector Charles Duelfer's exhaustive (thousand-page-long) report of October 6, 2004, which concluded that "Saddam Hussein did not produce or possess any weapons of mass destruction for more than a decade before the U.S.-led invasion."

Grave doubts on the Bush administration thrust toward war were expressed by the French and German governments, but, despite strong public opposition and even a serious split in his own party, it was strongly supported by English prime minister Tony Blair.

As American and British military forces began to assemble in Saudi Arabia, Kuwait, Qatar, and Turkey, the Iraqi government attempted to find means to stave off attack. Having no diplomatic channel, it explored other means. One demarche was made through the former head of the CIA's office of counterterrorism, Vincent Cannistraro. He reported that the Iraqis offered to allow several thousand American troops or FBI agents to scour the country to prove that they did not have weapons of mass destruction or delivery vehicles. According to Mr. Cannistraro, the Bush administration "killed" the Iraqi proposal.

In an effort to find out exactly what the Iraqi government

proposed, I went to Baghdad on February 1, 2003, to interview Deputy Prime Minister Tariq Aziz, whom I had first met twenty years before. In our two-hour session, I kept probing at what Iraq might do to head off disaster. Finally Aziz cut me off, saying that "America has long since decided to attack Iraq and nothing Iraq could do would prevent it."[*] At the same time, February 5, Secretary of State Colin Powell made a forceful high-tech presentation at the Security Council of the American case for war. Almost all of what he said has subsequently proved to be untrue, but it then swayed the Security Council and American public opinion. (Secretary Powell apologized in May 2004 for misleading the Security Council and the American public.)

The attack on Iraq began on March 20, 2003, as it had in 1991, with a furious air assault designed to "shock and awe" the Iraqis into a quick defeat. No exact figures will ever be known, but about ten thousand Iraqi civilians and tens of thousands of soldiers were killed in three weeks. British and American casualties were comparatively tiny: 128 American and 31 British soldiers. Most were killed by "friendly fire." Allied firepower was overwhelming. Among other devastating weapons some 13,000 "cluster munitions" exploded into 2 million cluster bombs, wiping out whole areas. The close coordination of ground and air forces and the disruption caused by intense bombardment before

---

[*] I had made clear that I was there as a private citizen and scholar, and went back to America to report to an audience largely composed of senior current and former government officials at the Johns Hopkins University School of Advanced International Affairs what had transpired, and to give my impression of Baghdad on the eve of the war.

the ground assault explain the contrast in casualties. As one military observer commented, "whole divisions were destroyed." Electronic warfare alternately disrupted Iraqi communications or used them to locate targets. In every sphere of activity, the Iraqis were outgunned, outnumbered, outclassed.

However, the advancing forces encountered pockets of resistance that fought desperately. Particularly in Basra, Baghdad, and Mosul, even when Iraqi military formations had lost all coherence, small groups fought on. Individual cases of Iraqi suicidal bravery were frequently reported—soldiers with rifles battling tanks and armored vehicles. But the Iraqis never had a chance. As a "war," the conflict was quickly over. By the middle of April, the Iraqi army had ceased to exist. On April 16, 2003, President Bush declared Iraq "liberated."

What happened then stunned the American and British commanders: having wiped out the Iraqi army, they were unable to declare victory. Peace did not come. The war had taken on a new shape. That new shape baffled the occupying forces and led to actions that as often fueled resistance as pacified the country. To understand what happened, it is necessary first to understand the context in which the allied soldiers operated and, second, to analyze the groups that inflicted more casualties on them than the war itself. First, the context.

Before the invasion, Iraq had lived for a decade under a grinding sanctions program that stripped away the savings of the new middle class and caused them to sell their possessions to buy food and clothing. Despite their desperation, there was little robbery or violence. For all its many terrible aspects, the Baath regime had suppressed crime and banned firearms. When I visited Iraq on the eve of the American invasion, it was possible to walk anywhere day or night in complete safety. That would dramatically change. Part of the reason was that Saddam in October 2002 amnestied scores of thousands of prisoners. Most were political prisoners, but thousands were common criminals, including some who were being held for murder, rape, and armed robbery. Then, in the chaos of the invasion, as the Iraqi army collapsed, they and many normally law-abiding but then distressed people broke into arsenals to seize supplies and weapons. As weapons were sold or passed out to relatives and friends, almost every Iraqi acquired an assault rifle; many acquired machine guns or even rocket launchers. The American Congress and public were not informed until eighteen months later, the end of October 2004, that a cache of 380 tons of powerful conventional explosives, particularly useful for car bombs and attacks on armored vehicles and aircraft, had been looted in April 2003. The ammunition dump, just thirty miles from Baghdad, which had been kept under seal by the International Atomic Energy Agency, was captured by American forces during the invasion but was then left unguarded. Failure to protect or destroy it certainly caused many American deaths.

At the same time, tens of thousands of soldiers deserted what was left of their units. Finally, on May 24, the newly appointed American administrator, L. Paul Bremer III, abruptly demobilized what was left of the army. Nearly half a million soldiers were told simply to go home. (Bremer's predecessor, General Jay Garner, had planned to keep them together to be used, and paid, as labor battalions.) The ragged, hungry, defeated soldiers took their weapons with them. When they arrived home, they found that no one had any money. Salaries were not being paid even to hospital staffs, but having no money made little difference since, in the runaway inflation, money was literally not worth the paper it was printed on. What mattered were things. Above all, food. Food was in such short supply that in April, May, and June of 2003, starvation was a clear and present danger. Because the bombing had destroyed purification and sewage facilities and the electricity to run them, even bottled water, the only kind that one could drink without getting sick, was too expensive for most people. Since everyone was frantic, looting became so common as almost to seem normal. Everyone took what he needed and defended what he could.

Neighbors who had managed to store up rice, dried beans, and flour gathered together, often under the direction of religious authorities, to protect their families and houses. The desperate poor, driven by hunger, and criminals, driven by greed, formed gangs that prowled the streets looking for targets. The police had drifted away from their stations to join either the vigilantes or the gangs. Iraqi cities became "free fire zones" with the stronger, better-armed, and more ruthless seizing control. Crime not only could not be investigated or punished but had lost even definition. Looting became a form of shopping. Industrial installations were stripped of machines, tools, copper wire, pipes, even light fixtures. Government buildings lost their windows and doors; Saddam's palaces, their erotic and gaudy art; Tariq Aziz's house, its library. The most tragic of the victims was the great Baghdad Museum of Antiquities, where thieves with chain saws cut apart ancient statues and scooped up thousands of priceless artifacts. A thriving market sprang up with foreign buyers feeding off local misery. That outrage was at least understandable; the burning of the National Library with its vast collection of ancient manuscripts was pure vandalism. American troops stood by and refused to intervene.*

---

* The State Department's Future of Iraq Working Group had prepared a list of sites to be surrounded and protected by American forces. The list was passed to Undersecretary Douglas Feith of the Defense Department, but he did not communicate it to the American military command. The Museum of Antiquities was number two on the list. See Peter W. Galbraith, "How to Get Out of Iraq," *The New York Review of Books*, May 13, 2004.

No provisions had been made to train administrators; only one senior official was proficient in Arabic. Whole cities were without police, firefighters, sanitation workers, or doctors. Apparently, the occupation authority administration had simply assumed that existing Iraqi personnel would continue to work, although no effort was made to help them do so. Such actions as were taken showed a stunning lack of sensitivity. Perhaps the worst was the reopening of the Abu Ghuraib prison, made notorious by Saddam's torturers and execution squads. Even the warden who had overseen the "disappearance" of thousands of Saddam's victims was reappointed by the American administration.

Out of this series of inactions, maladroit moves, and chaos arose the groups that would fight the American occupation forces during 2003 and 2004. Looking back, we can see a process: it began with looting versus protection; then increasingly large groups of people attempted to force the occupation authorities to disgorge food and other supplies by demonstrating. Tens of thousands of civilians took to the streets to demonstrate, peacefully for the most part, in Mosul, Falluja, Baghdad, and other cities. Seeing their demonstrations as a sort of rebellion, American forces began in April 2003 to fire on them. As casualties mounted, anger grew. That anger led to the first serious, military-style attack on U.S. forces on May 1 at Falluja, where American troops had just killed at least fifteen civilian protesters. Thereafter, month by month, demonstrations, protests, firings by troops upon crowds, raids on houses, and attacks on troops escalated.

At first, the American authorities dismissed the Iraqis as "Baathist remnants," but attacks grew in intensity, numbers, and distribution. By summer, they included attacks on installations that were thought to justify or solidify the occupation (such as the blowing up of an oil pipeline twice, on August 15 and 17, 2003) and on Iraqis who cooperated with the occupying force. Then even the

United Nations headquarters in Baghdad was blown up on August 19. Apparently those who carried out the attack* thought that the United Nations was providing cover for the occupation. In that attack, the head of mission, Sergio Vieira de Mello, and his chief of staff, Nadia Younes, were killed. Later, the embassy of Jordan, whose government was training a new American-sponsored army, was bombed. A member of the Iraq Governing Council was assassinated in September, and in December someone tried to kill the American administrator at the same time Saddam Husain was finally run to ground and captured.

In the early months, the attacks appeared isolated and unsystematic, but soon they began to be organized, by whom, no one yet knows. As attacks were coordinated, American officials blamed outside agitators, whom they loosely called al-Qaida. Secretary of State Colin Powell claimed, during a brief visit to Baghdad on September 14, 2003, that there were up to two thousand foreign militants then in Iraq. No evidence for this claim has appeared, and significant foreign intervention appears unlikely. Known al-Qaida militants are Sunni fundamentalists who would have to be housed, fed, and given mobility by local groups. Experience shows that the introduction of even a few foreigners into any society is a difficult operation, requiring long advance preparation and considerable organization. That did not exist in Iraq in 2003 and 2004. Moreover, outside agitation was not needed: the attacks on American and British forces usually came after clashes between troops and civilians or against Iraqis considered "quislings."

---

* Who actually did is unknown, but some group calling itself Armed Vanguards of the Second Muhammad Army issued a statement saying that it was responsible.

Initially, anti-American demonstrations were in predominantly Sunni Arab areas but, increasingly, the Shia Arabs became involved. On September 2, 2003, in defiance of a ban, tens of thousands Shiis marched through the holy city of Najaf. When troops tried to prevent them, tried to arrest the leaders, or fired on them, attacks grew in scale to assume the proportions of a guerrilla war. In October and November 2003, guerrillas managed to shoot down three helicopters; on December 9 at Mosul forty-one American soldiers were wounded; and on December 27, 2003, in Karbala, attacks involved car bombs, mortars, and machine guns. In such daring escapades, robbery was not the aim. What had begun to coalesce, although still without identified leadership, was a national rebellion involving, at the lowest estimate, at least five thousand combatants and, consequently, many times that many supporters. By October 2004, the estimated number of combatants and their active supporters had risen to about twenty thousand.

As American casualties mounted, the British and American governments had encouraged outside military participation. India was asked for a full army division, but, pointing out that its parliament had declared the war unjust and that thousands of Indian troops, under British imperial command, died conquering and occupying Iraq in the First World War, India declined. Small contingents came from other countries up to a total of about nineteen thousand. After suffering a terrorist attack in Madrid, Spain had a change of government and pulled out in April 2004.

To supplement or replace them, about four hundred of what are known as "private military firms" have flocked into the country so that much of the "security" tasks have become privatized. The largest provider is Halliburton, the company formerly headed by Vice President Dick Cheney and from which he still draws a substantial yearly payment. Growth in the number of mercenaries has been rapid. Blackwater USA, which was founded only in 1998, already has revenues of more than $1 billion. These companies are anxious not to be termed suppliers of mercenaries since mercenaries are illegal under the Geneva Conventions, but the distinction is difficult to draw. The numbers are huge: one firm claims to have more than ten thousand highly trained former soldiers on its rosters and another has about five hundred Gurkhas and the same number of Fijians now in Iraq. In total, these "troops" in Iraq number roughly twenty thousand or double the size of the British expeditionary force. They guard senior officials, patrol pipelines, and occasionally fight. Some have also been involved in the interrogation and torture of "enemy combatants"—a new term coined by the Bush administration to avoid having to accord the prisoners the status of prisoner of war as defined by the Geneva Conventions—at Abu Ghuraib and other prisons. They earn roughly three times the salaries of uniformed soldiers, but they are not subject to either military or civil jurisdiction and work only under the control of their employers. Under them as "shadow soldiers," the guerrilla war in Iraq has been partly outsourced.

Even with their help, the American military realized by the spring of 2004 that it could not handle the rebellion and began recreating Iraqi military and police formations. The aim of this policy is eventually to turn the war over to native military units as America did in the Vietnam War. In that war, America had inherited from the French a large army. In Iraq, of course, it has not, so it is trying to create one. The results have been disappointing. The idea that a local militia can accomplish what America's own powerful army cannot do is not policy but fantasy. True, in the days of their Iraqi empire, the British used such a force—composed of the Christian Assyrians—but only as auxiliaries to their army and air force. The Iraqi "Interim Government" has similarly used Kurds as auxiliaries to American forces. An Iraqi army is unlikely to fight insurgents with whom soldiers sympathize and among whom they have relatives. But, hoping that they would, American forces began to move out of Baghdad and other cities in April 2004. Then, realizing that their withdrawal had not caused the insurgents to stop fighting, they began in September a series of massive attacks on Shiis in Karbala and Najaf and on Sunnis in Falluja and Samarra, aiming to crush them before the elections scheduled for January 2005.

While administration officials had always maintained that the insurgency was a small-scale affair, involving only a few "die-hard Baathists," by July 2004, American intelligence officers conceded that at least 50 organizations comprising upwards of 20,000 combatants and active supporters were involved. They also affirmed that while some foreigners had entered Iraq to fight against the Americans and British, it was Iraqis rather than foreigners who were at the center of the resistance. The insurgency had become a national war being fought with guerrilla tactics.

Other recent or current wars have demonstrated that the prognosis for winning such a war is not good: in their war in Algeria the French employed over three times as many troops, nearly half a million, to fight roughly the same number of insurgents as America is now fighting in Iraq. They lost. After forty years of warfare against the Palestinians, the Israelis have achieved neither peace nor security. Both tsarist and Communist Russia have been fighting the Chechens since about 1731. President Putin's Russia is still at it with no end in sight. The commander of the U.S. First Infantry Division, Major General John Batiste, commented, as his predecessors had found in Vietnam, that such wars "cannot be won militarily."

Can they be won politically?

That question was not really posed until well after the invasion. So stunning was the lack of preparation that the ensuing chaos seems almost deliberate. To understand this dimension of the war in Iraq, it is necessary to go back to the first days of the spring of 2003. While the Department of State had developed a comprehensive plan, it was simply shelved by the men in control of the Department of Defense, led by Secretary Donald Rumsfeld and the Neo-Conservative clique he had assembled under Undersecretaries Paul Wolfowitz and Douglas Feith. They wanted a "muscular" approach to the occupation in line with their declared objective of reconstructing much of the world in America's image. They did not seek transition. Their approach was adopted because they had the personnel, facilities, and means of transport firmly in hand and because President Bush authorized the appointment of a man in agreement with them, a retired general who had become a defense contractor, as the American "proconsul." General Jay Garner arrived in Baghdad on April 21, 2003.

Garner brought with him the nucleus of an Iraqi advisory committee. The key figure was the Pentagon's favorite, Ahmad Chalabi, an Iraqi who had left Iraq as a teenager and who had been convicted of fraud while a banker in Jordan. The Pentagon flew to Iraq Chalabi's armed militia, much as the British had arranged that the man they installed as king in 1920 arrive with his. Chalabi would have less luck than King Faisal. He would later be accused of defrauding the American government of millions of dollars and revealing its communications secrets to a foreign power.

Garner also enrolled the rival leaders of the Kurds, Masud Barzani and Jalal at-Talabani, along with the man who was destined a year later to emerge supreme, Iyad al-Allawi, former head of the CIA-sponsored anti-Saddam terrorist group, the Accord (Arabic: *al-Wifaq*). Except for the two Kurds, Garner's advisers were virtually unknown in Iraq. That was the problem the British had faced with the then almost unknown Faisal in 1920. Meanwhile, domestic opposition groups sounded the alarm, drawn from their memories of "British Iraq," that the occupation "interim" authority might turn into a permanent mandate or colony.

General Garner seemed unable to grasp what was happening and tried, simplistically, to solve political problems with military force. His brief time in office was a month of chaos. On May 7 he was replaced by L. Paul Bremer III. A retired foreign service officer, Mr. Bremer had served as ambassador-at-large for counter-terrorism and subsequently as head of the National Commission on Terrorism. Between those times in government he was managing director of Kissinger Associates. What initially recommended him for his post in Iraq was that, as a strong supporter of the Neo-Conservatives, he was enthusiastically endorsed by Vice President Dick Cheney and Secretaries Rumsfeld and Wolfowitz; as he was a former foreign service officer, the State Department could not object; and because of his former government positions and writings he symbolized a link, which the administration was actively publicizing, between terrorism and Iraq. President Bush described him as a "can do" man who "goes with the full blessings of this administration." He was greeted with less enthusiasm by human rights groups, as he had advocated recruitment of foreign agents with "dirty" records. Like Mr. Wolfowitz, he had reacted to the al-Qaida attack on September 11, 2001, with a call for war on Iraq. Unlike the hapless General Garner, he was being sent to Iraq with a plan.

The plan, presented to the Security Council on May 9, named America and Britain as "occupying powers." Exactly what that meant was not clear. In one of his first statements, Mr. Bremer said, "We are not here as a colonial power. . . . We are here to turn over [authority] to the Iraqi people as quickly as possible." But Mr. Bremer moved slowly to involve Iraqis. He initially followed the path General Garner had marked out with an advisory "political council." But after several critically important groups of Iraqis refused to work with it, Mr. Bremer decided by July 2003 to change its name if not its role to "governing council." To it, he appointed twenty-five Iraqis: thirteen Shia Arabs, five Sunni Arabs, five from the rival Kurdish factions, one Turkmen, and one Assyrian Christian. Included among them were the leader of the Iraq Communist Party and three women. They held their first meeting on July 13, 2003; their assigned tasks under American supervision were to prepare a budget and approve a constitution.

Meanwhile, the occupation authorities had begun to reconstitute administration with some 250 town councils in the countryside. Pragmatically, they often reappointed local officials who had long served the Saddam regime. Because most of these men were unpopular, they were dependent on the occupying authority to protect them and so could be relied on not to foment opposition to it. However, their involvement inevitably made the aim of promoting representative government more difficult to achieve and sometimes provoked riots. To be fair, the authorities had little choice. Whereas the large numbers of American occupation authorities in Germany and Japan in 1945 had been trained in their languages, very few could communicate with natives in Iraq. Inability to understand what was being said caused constant misunderstanding, much anger on both sides, and a number of deaths of Iraqis.

After months of discussion, on March 8, 2004, the Governing Council approved the interim constitution. Officially known as the "Transitional Administrative Law," it was written by American lawyers and was seen by only a few Iraqis before it was promulgated. It was proclaimed by Secretary of State Colin Powell as "a major achievement" and by British prime minister Tony Blair as the "foundation stone" of the new Iraq. Many Iraqis saw history repeated: it was almost exactly eighty years before, in 1924, that British officials similarly advised and guided a committee of carefully selected Iraqis to write a constitution. It too had proclaimed democracy. On paper, the phrases rang with eloquence, but they were not grounded in reality. When it was needed, ambitious Iraqis simply paid it no heed: in roughly thirty years, Iraq suffered under actual or disguised dictatorships through a dozen coups, of which the last brought the Baath to power.

The constitution did offer at least verbally a vision of a new democratic Iraq with an independent judiciary and, most important of all, civilian control of the military. It had been the imbalance between military power and civil institutions that decade after decade had wounded Iraq, but the constitution could not rectify that balance. Despite the historical record that so clearly showed that the army was the enemy rather than the protector of Iraqi freedom, the Bremer administration immediately set out creating a new military force before compensatory public groups could take root. Moreover, like the 1924 constitution, the 2004 constitution evinced little trust in the people. It specified national elections to be held by January 31, 2005, for a 275-member assembly that would then choose a president and two deputy presidents. These three would, in turn, select a prime minister who would wield executive power.

Speaking for the Iraqi majority Shia community, its most important leader, Grand Ayatollah Ali as-Sistani, professed to be outraged. Not only were the Americans selecting the government without recourse to the people, but they were engaging in other policies that were illegal in international law and that undermined the prospects for Iraqi sovereignty. (I will discuss these in the following section.) He refused to meet with Mr. Bremer to discuss the constitution and warned that it "enjoys no support among the Iraqi people." Although less outspoken, the Kurdish leaders appeared to agree that the constitution was either irrelevant or deeply flawed. Opposition was expressed violently when a car bomb killed the head of the council on May 17; two weeks later gunmen tried to kill another council member, and on July 14 they killed the American-appointed governor of Mosul. At base, the problem was not in the document but in the lack of consensus: the Shiis were pressing for Iraq to be an Islamic state while the Kurds were pressing for at least federal status. Both they and the Sunni Arabs feared that it did not assure them of complete independence. They need not have worried. It was the shortest-lived constitution ever promulgated. By international law, it lapsed when the occupation authorities turned over political leadership to the Iraqi Interim Government.

"Independence" evoked deep memories for Iraqis. Like the Americans in 2004, so the British in 1924 had proclaimed that the constitution was the first step toward self-determination. But, Iraqis remembered, the British ruled Iraq overtly or covertly for the next thirty-four years. Would the Americans do the same? Iraqis asked. They watched American firms getting contracts worth billions of dollars that turned some from near bankruptcy to high profit. In "independence," would Americans give up their virtual economic monopoly? Iraq was also known to have the largest pool of untapped oil in the world. Could America afford to leave? Would the Western powers not prefer to work, as they had in the past, with undemocratic and unpopular governments that would not challenge these interests? Was 2004 just an interlude between dictators?

Many thought it was better not to ask. With the help of a United Nations envoy, on June 1 the Iraqi Interim Government replaced the Iraqi Governing Council. The power of the new government was to be sharply limited—its armed forces would remain under operational control of the American military; its finances would similarly be overseen by American officials; it would have no authority to amend edicts from the American occupation or even to enact new laws; and its key ministries would be dominated by American-appointed commissions. As one of the new ministers, Haider al-Abadi, commented, his ministry will actually be run by a commission whose members were selected by Mr. Bremer and will hold office for five-year terms, far beyond the planned eighteen-month tenure of the interim Iraqi government. Minister Abadi went on to say, that he regarded the transition as meaningless: "If it's a sovereign Iraqi government that can't change laws or make decisions, we haven't gained anything."*

It was at this point that the American proconsul, Paul Bremer, slipped out of the country secretly on a military jet. What did he leave behind? Consider the following answer.

Week after week and month after month for a long time to come we shall have a continuance of this miserable, wasteful, sporadic warfare, marked from time to time certainly by minor disasters and cuttings off of troops and agents, and

---

* As a sign of how little the minister was regarded, he commented that "No one from the U.S. even found time to call and tell me" of Bremer's edict; he learned about his actual demotion in the press. On this episode see Yochi Dreazen and Christopher Cooper: "Digging In: Tight U.S. Grip will Guide Iraq Even After Handover." *The Wall Street Journal*, May 13, 2004.

SIX

WHOSE IRAQ?

Whose country is Iraq going to be? American in some form of hegemony under an American-appointed Iraqi leader? Shia under an Islamic fundamentalist government? Sunni Arab in a secular regime? A "guided democracy" (read: dictatorship) under a military junta or a single strongman? A United Nations "mandate"? One country, two countries, or three? All of these are possible. Which is the most likely will be determined by the long reach of Iraqi experience and by trends that have begun to take shape under the American occupation. Attempting to see ahead is the purpose of this concluding part to *Understanding Iraq*.

Much of this book has been devoted to illustrating and explaining the long reach of the Iraqi experience. The record shows that for thousands of years Iraq has been intermittently a rich and creative society, but that its bursts of great cultural efflorescence have been interspersed with devastating tragedies—foreign invasions, massive destruction, domestic tyrannies, expulsions of population, plagues, famines, and genocide. The Iraqi people have shown great resilience, but over the past century they have been worn and battered by an almost constant onslaught. Iraq today is a wounded society.

Over its long history, the one group that has seldom "owned" Iraq was its people. Perhaps the last time they did so was thousands of years ago in Ubaidian times. Thereafter, when pressed by economic, social, or military threat, they often sought shortcuts to safety. In the past, they have often been led along these shortcuts by dictators, *lugals*, great men, "kings of totality," "sole leaders," "hero-presidents," whose guidance has come at great cost. Although the trails they blazed rarely led to safety and prosperity, their calls were hard to resist. Often the Iraqis had no choice. They were bludgeoned, bullied, or tricked by local strongmen or foreign invaders into following. Over time, following became a habit. So the challenge of Iraqis today is to break the chains of their history, find their voice and establish their claim to govern themselves in their own best interests. Can they do it? They are a hardy, proud, and capable people, but the choice is not completely theirs.

It is in the context of this challenge that the actions of occupation authorities must be weighed. Have they encouraged movement toward an Iraqi society that is open, free, and peaceful, or have their policies worked against these objectives? Will observers a generation from now look back on this period as merely an interregnum between dictators? Before evaluating the effect of American programs in Iraq itself, let us consider what has become a pattern of American action over the last half century.

The American policy designed by the ideologically-driven group known as the Neo-Conservatives proclaimed that

America had the right, indeed the obligation, to impose its way of life on the whole world. Iraq was an early step in what was to be a new "crusade," which would be accomplished by warfare, essentially unending and everywhere. If remaking the world in America's image is really the Neo-Conservatives' aim and warfare is the means they intend to adopt, they are poor students of history. The results of American military interventions show a different pattern: Guatemala in 1954, 1966, and 1972; Lebanon, 1958; South Vietnam, 1960s; Republic of the Congo, 1967; Nicaragua, 1978 and 1982; Grenada, 1983; Panama, 1989; Iraq, 1991; Somalia, 1993; and Afghanistan, 2001 to name a few of the thirty-five American interventions since the Second World War. What happened? No undemocratic countries turned into democracies. Establishment of democracy happens indigenously or it does not happen at all. It has never been imposed on governments or nations at the point of a bayonet.

But the bayonet can make less likely or even impossible the establishment of a more representative, more tolerant, and more peaceful society. Consider also that side of the record. The CIA (with help from the British intelligence agency MI-6) in 1953 organized a coup d'état to overthrow the democratically elected Iranian government of Mohammed Mossadeq. This action returned to power the undemocratic government of the shah, led to the revolution that established the Islamic fundamentalist government, and devalued the "image" of America as the patron of democracy.

The relevance of this to Iraq is twofold: first, partly because of it, Middle Easterners believe that when the United States disapproves of a government, it moves to "destabilize" it; and, second, the initial success of the coup in Iran, considered a classic case of espionage, set a style that America has followed in several other countries,[*] including Iraq. In Iraq, the CIA gave covert espionage assistance to the Iraqi Baath to enable it to come to power in a coup d'état. As in Iran, it took this step because of its disagreement with an existing government. The government of General Abdul Karim Qasim was not democratic, but, as in Iran, the coup started a process that was to lead to a far worse government. America was also complicit in the ensuing bloody purge of members of the deposed Iraqi government. These events convinced Iraqis that while America talks of democracy, it sometimes behaves like the Mafia. Some ambitious Iraqis also concluded that espionage is "politically correct."

What was politically incorrect was to oppose the desires of the Great Powers. Prime Minister Mossadeq in Iran tampered with oil; Prime Minister Qasim in Iraq flirted briefly with the communists. They had to go. Thus, when outsiders sneer at Middle Eastern paranoia, Iraqis, Iranians, and others speak of history.

Following the coup that opened the way, Saddam Husain built the Baath Party and took power. When he was firmly ensconced, the Reagan and first Bush administrations found reason to associate the United States with him. Even though the Saddam regime's record of tyranny, torture, and slaughter was already evident to Iraqis and well known throughout

---

[*] U.S. president Dwight Eisenhower and British prime minister Harold Macmillan in 1957 approved a joint CIA–MI-6 plan to encourage "incidents" within Syria, then stage a fake frontier battle as an excuse for invasion after which the leaders of the Syrian government would be "eliminated." The coup was planned by CIA assistant director for the Middle East Kermit Roosevelt, who had engineered the coup against Mossadeq in Iran. It was not carried out because the governments asked to participate, Jordan and Iraq, refused. Documents on the plan were published by *The Guardian* on September 27, 2003.

the world, they sent presidential emissaries (notably Donald Rumsfeld) on highly publicized visits to assure Saddam Husain that America supported him. In addition to this diplomatic support, they gave or lent his regime money and weapons and provided highly sensitive intelligence information that enabled it to defeat Iran. Despite their public condemnation of such activities, they also facilitated the sale to Iraq of feedstock for chemical and biological weapons and the equipment to manufacture them—even when they knew that these weapons were being used against Iraqi civilians. They also encouraged or allowed corporations to sell components for nuclear weapons.

At the conclusion of the Iraq-Iran war in 1988, the first Bush administration indicated no opposition to Saddam Husain's declared intention to "rectify" Iraq's frontier with Kuwait even though such action might involve the use of military force. When the Iraqis went further than American officials anticipated—endangering not American principles but American interest in oil and money—America went to war. Then, in 1991, having defeated Iraq, it did not prevent the regime from savagely suppressing those Iraqis who were trying to free themselves. America allowed Saddam to use helicopter gunships against Shia insurgents, pulled its forces back to allow Iraqi police and Republican Guard forces to move against them, and prevented the rebels from arming themselves. As a result, thousands of Shiis were slaughtered. The survivors believe that the United States took these actions, at whatever cost to the well-being and freedom of Iraqis, solely to block a possible extension of Iranian influence. Their belief is partly borne out by the contrast of American policy in the Kurdish north, where Iranian intervention was not a danger. There, the United States protected the rebel movements. Then, using Kurdistan as a secure base, America began covertly to support terrorist groups attempting to overthrow the Baath regime and to murder its leaders. Thus, whether helping to create the regime of Saddam Husain or protecting it against an external enemy, or attempting to destroy it, America promoted authoritarian, subversive, and illegal movements and individuals. Such "dirty tricks" are not the road toward what must be the long-term American objective of creating a peaceful, reasonably tolerant, and popular Iraqi government.

So what has America done since its invasion of Iraq in 2003, and how should these actions be judged by its aims and interests?

Events and tendencies to date are not encouraging. In the aftermath of the destruction of the tyranny of Saddam Husain, America has convinced many, probably most, Iraqis that it is not occupying their country to promote freedom but to engage in a new and more sophisticated form of imperialism. A recent independent public opinion poll holds that only 2 percent of Iraqi Arabs view the United States as liberators. This disbelief has evoked from them an outburst of nationalist ardor that today fuels a national uprising: during the summer of 2004 American forces were attacked by insurgents at least sixty times a day. "If we stay anywhere more than five minutes, they start shooting at us," said an officer in the First Cavalry Division. If Americans remembered their own history, they would not be surprised. Writing about another insurgency, the American Revolution, the English statesman Edmund Burke commented in 1775 that "The use of force alone is but *temporary*. It may subdue for a moment; but it does not remove the necessity of subduing again: and a nation is not governed, which is perpetually to be conquered." As costly in lives and property as the Iraqi uprising has been—with more than fifteen thousand civilian deaths (and some estimates at several times that number) and about ten thousand Iraqi captives held in American prisons—it has created three new trends that will shape Iraq's future.

The first of these trends is that opposing foreign occupation has at least temporarily driven Sunni and Shia Iraqi Arabs together in common cause. That happened briefly in their opposition to the British in 1920, as then colonial secretary Winston Churchill commented, but under British occupation they were quickly driven apart. Sunnis were favored and Shiis were pushed out of participation in government. In 2004 the two communities are working together or at least in parallel against a common foe, the American occupation. Understandably, the Kurds have stood aloof from this nationalist struggle.

Despite their often bitter internal divisions, the Kurds have long aspired to independent statehood, and they have come close to that dream during recent years. Profiting from aid funds and from trade with Turkey, Iran, Syria, and Arab Iraq, they have made considerable economic progress. This has encouraged the vast majority of them to petition for a vote on independence and to deck their country with Kurdish flags. They have now more or less unified the various guerrilla organizations that fought against the Baath and often against one another into something like a national army that has become the only effective indigenous military force in Iraq. Their participation in whatever Iraq emerges will always be partial, but the dangers they face from Turkey and Iran will make affiliation the least unattractive of their current options. These quite different forces—Arab opposition to foreign rule and Kurdish fear of foreign intervention—will cause Iraq to hang together as a single state, although it will probably be forced to acknowledge its deep schisms by becoming federal. Going further than federalism, attempting to "balkanize" Iraq is likely to turn it into an eastern Balkans—a maelstrom of ethnic groups. At the minimum, splitting Iraq into pieces will provoke flights of ethnic or religious groups from one area to another, disrupt public services, hamper trade, cause massive human rights abuses, and prevent healing of the wounds of the Saddam era.

National uprisings against foreign occupation, the second trend, strip away from both sides the thin veneer of civility that separates us all from the bestial. If the struggle goes on long enough, acquisition of the habit of violence causes a society to become unhinged: its basic institutions cease to function, neighbors fall apart, even families lose coherence, and the customary lines that separate acceptable behavior from crime are erased. Then whole societies falter. This happened in Algeria in the 1950s and early 1960s, and in Vietnam in the 1960s and early 1970s, and is happening today in Afghanistan, Kashmir, Çeçneya, and a dozen other countries. Depending on how long the struggle in Iraq continues and how violent it is, Iraqis could be dragged down into a kind of social incoherence from which they are likely to give up trying to attain a just and peaceful society. In such circumstances the rise of "warlords" (as in Afghanistan) or a new dictator (the "ghost of Saddam") becomes almost inevitable. Thus, the very fact of American military involvement in Iraq accentuates tendencies that America has announced it wishes to avoid.

The third trend is the American quest for security. Understandably, the occupation authority put its aim of security into conflict with the Iraqi aim of "sovereignty." (Most Iraqis with whom I have talked do not regard the current interim government as more than an American puppet and do not consider it as having solved the issue of sovereignty.) Those Iraqis who aspire to complete sovereignty are prepared to create complete insecurity. They do this not only by fighting against American troops—and the scale of this warfare is far more than most outsiders realize—but also by making difficult or impossible rebuilding efforts and even by destroying the infrastructure upon which their future well-being depends. The stronger the repressive measures employed to create "security," the more desperate becomes the struggle for sovereignty; and the more desperate that becomes, the more the society is moved away from civility and security toward brutality and chaos. This is true because in clandestine struggle, the survivors are likely to be those who are tightly organized under authoritarian leadership.

Apologists point to the fact that the occupation authorities moved progressively and with all deliberate speed from a "political council" to a "governing council" to an "interim authority" to an appointed assembly that approved American-appointed and American-"advised" ministers under an American-chosen prime minister. True, the latest stage in the process has left large "reserved" areas of government in American hands—just as the British did in the 1920s. Americans will exercise ultimate control over the military, finances, oil, and foreign affairs and will continue to influence the choice of senior officials. But, on paper, the record, given the circumstances of occupation, has a certain coherence, even a certain validity. However, it has two fatal flaws: the first is that it was all done *by* foreigners *to* Iraqis, and the second is that it started from the *top down* rather than from the *bottom up*.

Consider first the Iraqi personnel of the American-installed government. In placing its own agents in office, regardless of their standing among Iraqis, the occupation authority followed the path of the British in the 1920s. Like Faisal, whom the British then made the king of Iraq, so Ahmad Chalabi and Iyad al-Allawi were paid American agents. Faisal had never lived in Iraq; Chalabi had been out of the country since he was thirteen; and Iyad al-Allawi had been abroad for about thirty-five years, longer than most Iraqis had been alive. Chalabi was the favored candidate of the Neo-Conservative clique that had designed the Bush administration policy toward Iraq; reluctantly, they dropped him. They did so not when his criminal financial background became known—it had been known long before the invasion —*or* when he was unable to account for the more than $30 million he was paid by the American government, *or* when it became known that he was using his access to American authorities to enrich himself from defenseless Iraqis, but when he was suspected of revealing American code breaking to Iran and to be engaged in counterfeiting money.

With him out of the way, the occupation authority focused on Iyad al-Allawi. According to independent public opinion polls in April 2004, he was the most disliked figure in the governing council. A former senior Baathist, allegedly an "enforcer ... involved in dirty stuff... with blood on his hands" in Saddam's repressive secret police (according to a well-informed former senior CIA officer), he broke with Saddam in the late 1970s, and Iraqi agents tried to kill him. Then, in revenge, operating out of London and Kurdistan, he led a CIA-subsidized group, the Accord (Arabic: *al-Wifaq*), in anti-Saddam terrorist attacks, one of which was alleged to have blown up a schoolbus full of children. Chosen ostensibly by the UN representative, former Algerian foreign minister Lakhtar Brahimi, but actually by the occupation authority as prime minister, he quickly exhibited his widely known violent and authoritarian character: six days after taking office, with American approval or at least acquiescence, he promulgated a law giving him power to impose curfews, restrict domestic and foreign travel, ban groups he deemed seditious, and order the detention of people he suspected to be risks to security. The new edict also empowered him to override civilian government by appointing "commanders" to administer areas of unrest. (That is, effectively, the whole of Iraq.) He proclaimed that "We will not allow some people to hide behind the slogan of freedom of the press and media," and on September 5, 2004, ordered his security officers to break into the Baghdad bureau of his most effective media critic, the radio and TV network al-Jazeera, and closed it down "indefinitely." Perhaps as important, he has created a new "supreme council for oil and gas," of which he is chairman, to approve contracts with foreign companies to exploit Iraqi energy. (That is, to control the most important sector of the Iraqi economy.) In short, his centralization of power is on a Saddam-like scale.

At the same time, he announced a crackdown on crime and toured Baghdad's police stations to assure the reinstalled Baath regime police that the government would support them against charges of torture or extra-judicial killing of prisoners. To convince the police that he was serious in his intent to exercise this power, the prime minister was reported by a respected Australian newsman[*] to have personally executed six handcuffed and blindfolded prisoners.

---

[*] Paul McGeough in *Sydney Morning Herald*, July 17, 2004. His report was based on interviews with witnesses who approved of what they reported, but were denied by Mr. Allawi's office. While possibly untrue, the story is widely believed in Iraq because it is in character with what is known of Mr. Allawi's past as a secret police agent under Saddam Husain.

Consider, second, the occupation authority's institutional emphasis. Almost all its attention was placed on what might be termed the upper reaches of government, the various councils and the cabinet, and upon the codification of these institutions in the now defunct American-drafted constitution ("the Transitional Administrative Law"). If the American administration was trying to stabilize a new, reasonably free, and reasonably democratic Iraq, it began at the wrong end of the process. What the occupation officials should have remembered from their own experience at home in America is that what actually makes representative government work is neither written constitutions nor lofty offices nor even reasonably honest officials, but the participation of the citizens at the grass roots. Only when the people take charge of their mundane problems do they acquire the habits, skills, and self-assurance that make them able to restrain or guide government.

Although apparently the occupation authorities did not know it, Iraq has an old tradition of neighborhood self-government: neighborhods traditionally maintained their own churches, synagogues, and mosques; local watchmen ensured local "security" while respected neighbors prevented excessive disputes through arbitration and consensus. At the beginning of "British Iraq," neighborhoods also ran their own schools and hospitals. True, these tasks were often not very well performed because the society was then poor and ill-educated. So, in the quest for modernization and control, the British sought to replace this primitive "participatory" democracy with a centralized system. Centralization and modernization in-creased in the 1930s and became the norm in "Revolutionary Iraq." Despite centralization but because of modernization, professional men and women—engineers, lawyers, teachers, and doctors—created a more advanced form of participatory politics by forming professional unions (Arabic: *naqabat muhanniya*) to monitor government activities and influence policies. Under the Baath dictatorship, these organizations were partly co-opted and twisted to Saddam Husain's purposes. Thus, both professional and local traditions have withered, but they have not died. The Occupation authorities paid little or no attention to them, but from these withered roots could grow genuine popular participation in government. Indeed, they are probably the only hope for some form of representative Iraqi government. Democracy must be rooted, as Thomas Jefferson would have said, in the "soil" of Iraq if it is to grow. Very few plants and certainly not democracy grow from the top down. What the American authorities have done is thus the very reverse of what Iraqis needed. They focused on the rulers and neglected the people. Drawing up constitutions and appointing councils will prove barren in "American Iraq" as it did in "British Iraq."

Perhaps even more important, in its quest for "security," the occupation government put its money, authority, and attention on recreating those instruments of repression that have so often and so seriously harmed Iraq in the past. The most glaring example is, of course, the continuation of Saddam's prisons. Using the notorious Abu Ghuraib prison, and—with almost unbelievable insensitivity—even (under General Garner) appointing Saddam's chief torturer to continue to run it and (under Paul Bremer) engaging in dreadful forms of torture and murder at the Abu Ghuraib and other prisons was, to put it mildly, a ghastly blunder. After having rightly condemned the appalling attacks on civil liberties of the Saddam regime, America "at the highest level" has been exposed as having condoned and probably authorized torture. While in American detention, an undisclosed number of prisoners were permanently disabled and at least twenty-five were killed in 2003 and 2004. When accounts and photographs of torture, enforced degrading sexual acts, and other forms of humiliation circulated,[*] Iraqis asked themselves whether there was any *qualitative* difference between Saddam's dictatorship and American democracy; many concluded that "democracy," as Americans practiced it, was little different from tyranny, as Saddam had practiced it. Thus, the very concept of democracy was also a victim.

Equally maladroit, but ultimately even more dangerous, has been American policy on the Iraqi army. Initially, General Garner proposed to convert those Iraqi military units that were still embodied after the invasion into a labor corps. They could have carried out emergency repairs and been paid for their work. That was a reasonable idea and would probably have worked. However, Garner's replacement, Paul Bremer, reversed this policy and dismissed hundreds of thousands of soldiers, sending them home, ragged, hungry, and broke—but allowing them to keep their weapons. In desperation or greed, many took

---

[*] Brutal American practices at Abu Ghuraib and other American-administered prisons have been condemned by the Red Cross as "a violation of the Geneva conventions . . . which the Bush administration has said it regarded as 'fully applicable' to all prisoners held by the United States in Iraq." The Red Cross, after noting that some prisoners were hidden from it, issued its report in October 2003, about nine months *before* the scandal became public. After his investigation in the spring of 2004, U.S. Army Major General Antonio Taguba also described the practices as "in violation of international law and U.S. army doctrine." These and other reports were analyzed in an excellent article by Seymour M. Hersh in the *New Yorker* on May 10, 2004.

Torture was declared illegal in the 1994 UN Convention Against Torture, but U.S. Department of Justice officials in 2002 argued that the president could authorize "a wide array of coercive interrogation methods" without violating international treaties or the federal anti-torture law: "Coercive methods should not be considered 'torture'" unless they caused "organ failure, impairment of bodily functions or even death." White House Counsel Alberto R. Gonzales held that the Geneva conventions did not

to crime. Others formed the armed fist of the anti-American nationalist movement. For Bremer's policy, the American army has paid in blood. Bad as that policy was, much worse (at least for the Iraqi future) was also then set in motion. Faced with criticism for incurring increasing American casualties, with more than a thousand dead and perhaps ten thousand wounded, the American authorities decided to rebuild the Iraqi military and security forces. The aim was to use "tame" Iraqis to fight "wild" Iraqis. This is a policy that successively the British and various Iraqi dictatorships, including that of Saddam Husain, employed. The Americans thought that what Iraqi soldiers *could do* more effectively than Americans was to confront Iraqi nationalists. Not surprisingly, they have shown little willingness to fight their fellow countrymen. Many deserted, others refused to fight, and some joined the insurgents.[*]

As the histories of "British Iraq" and "Revolutionary Iraq"

apply to more than nine thousand prisoners held without charge as of May 2004 without access to legal counsel, chance of impartial hearing, and protection against inhuman treatment. No public record exists of even who they are. Many were kidnapped in third countries; others were shipped to places where it was known they would be tortured or "disappeared." Some were "interrogated" by private contractors, of whom some were not American citizens, and none of whom were under legal control. As the executive director of Human Rights Watch said, "The courts have ruled most of these techniques illegal."

[*] After assuming office, Prime Minister al-Allawi resumed an old British practice. Whereas the British used Assyrian Christian "levies" to fight Arabs, al-Allawi is using Kurds to attack rebellious cities such as Najaf, Karbala, Baquba, and Falluja. Such moves, of course, accentuate divisions between the Iraqi communities. The British move provoked a massacre of Assyrians in 1932; what al-Allawi's actions will do to stimulate resentment among Kurds and Arabs remains to be seen, but it almost certainly will.

show, no civilian government can long exist when the military is turned inward to domestic affairs. In the years since the middle 1930s, the Iraqi military overthrew governments more than a dozen times. Accentuating instruments of repression thus has made more likely the reversion to dictatorship.

What keeps the military in check in democratic countries is the existence of balancing institutions and customs. In America, they are investigated by the press and operate under civilian control and under law; in Iraq none of these is fully operational. The occupation authority carried on a vigorous public relations campaign, restricted the access to news of foreign reporters, and tried to shield itself from criticism by attacking the critics. Foremost among these critics was the Qatar-based TV and radio network known as al-Jazeera. Its local correspondents were harassed, its offices were attacked, and even the American secretary of state tried to convince the ruler of Qatar to close the station.

As most Americans would agree, a strong judicial system is at the heart of a democratic government. It follows that American policy should at least not weaken moves in Iraq toward achieving a rule of law, yet that is precisely what has happened. In short, a reconstituted Iraqi army is at best irrelevant; at worst, it could be the avenue to power of the ghost of Saddam Husain.

The Bush administration announced in 2001 that its fear that Iraq had weapons of mass destruction and was supporting international terrorism were the reasons for the invasion. As we now know, Iraq did not have those weapons and was not supporting the al-Qaida movement of Usama bin Ladin. So consider the fallout of the Bush administration policy in Iraq.

Events over the half century of the nuclear era show us a pattern. In fear of one another, governments have sought the weapons of mass destruction that they regarded as vital to their security. America acquired them for use against Japan and for deterrence against Russia; the Russians acquired them to deter America; the Chinese decided that they had to have them to balance Russia; the Indians, against the Chinese, the Pakistanis, against the Indians; the Israelis, against the Arabs; and now the Iranians and North Koreans, against America. Ironically, the American invasion, ostensibly made to stop the spread, may lead other nations—and perhaps eventually a reconstituted Iraq—to seek to acquire weapons of mass destruction out of fear of America.

The only way to avoid the spread of such weapons is collective banning of them. That would be in everyone's interest. Weapons of mass destruction are far too expensive to acquire and far too dangerous to keep. They create insecurity rather than security. Achieving at minimum an area-wide free zone will be very difficult and will require lengthy negotiation. America has begun the process in Iraq. If it stops there, it will ultimately fail. Whether it moves toward regional disarmament or not will do much to shape the Iraq of the future.

The Bush administration also charged that Saddam Husain's regime was allied to Usama bin Laden's terrorists. Every "proof" of this charge turned out to be bogus. Moreover, the connection was always inherently unlikely since, for all of its terrible faults, Saddam's regime was committed to secularism while Usama bin Ladin and his followers were religious fundamentalists. Illustrating this incompatibility, Usama denounced Saddam with the strongest curse in his vocabulary as an "infidel" (Arabic: *kafir*). But, in the aftermath of the American invasion, as Egyptian President Hosni Mubarak rightly said, the

Bush administration has created a hundred bin Ladens and made at least some Iraqis receptive to them.

So, both in terms of the "delicate balance of terror" (as one of the Neo-Conservative mentors put the nuclear issue) and in terms of terrorism (the centerpiece of the Bush administration), American policy has been self-defeating. It has created an entirely new form of instability for Iraq and greatly increased danger for America.

Many Iraqis believe that the underlying purpose of the American invasion of Iraq was not fear of Iraqi danger to America but greed for its oil. It is true that getting Middle Eastern oil on acceptable terms has ranked as one of the three or four central objectives of American administrations of both the Democrats and Republicans for the last half century. Seeking that objective is certain to continue. Whether or not controlling Iraqi production was a major reason for invasion, it is undeniable that American policy on oil will play a major role in shaping Iraq. So everyone concerned with Iraq—or the Western economy—needs to understand precisely what is at stake.

From their long experience with foreign exploiters, and with their recent memory of "crude oil politics," Iraqis almost certainly will react against "oil imperialism" with unending hostility. Thus, what happens to oil is critical to security in Iraq. Already, attacks on pipelines and other facilities underline the Iraqi willingness to use what the Israelis termed, for themselves, "the Samson Option": the willingness to pull down the temple rather than lose it to the enemy.

If America puts its access to Iraqi petroleum at risk by seeking to dominate Iraqi production, it has misunderstood what really is at issue. Getting oil on acceptable terms is not the same thing as owning or controlling the fields where it is produced or even setting the terms of sale. As the Arabian American Oil Company (ARAMCO) realized already in the 1930s and as the United States and Britain belatedly should have learned from the Iranian ("Abadan") crisis of the 1950s, which flag flies over a field is not crucial. What matters is that oil flows and that the price is acceptable. These two objectives need not conflict with Iraqi national integrity. In today's world, they can be almost automatically achieved. The nation in which the field is located and the purchasers share an interest in moving the oil. The nation earns nothing if it does not sell its oil, and, if it prices its oil above the world market, purchasers can go elsewhere. Thus, in oil as in other commodities, the market is largely self-regulating. Where this market mechanism broke down in the past was where there was a monopoly. The Iraq Petroleum Company (IPC) used its monopoly to regulate *both* price and volume of production. Under the occupation, the market mechanism has also been prevented from working: despite the Bush administration's proclaimed commitment to a free market, the occupation authority sold Iraqi oil below world prices to British and American oil firms by fiat. Feeling that they have been exploited and that oil is what keeps Americans in Iraq, Iraqis can do little to oppose that policy except to commit acts of sabotage, which the insurgents are now doing; when the Americans leave, those attacks will cease. Then, presumably, oil will move freely under conditions of the world market.

Other than acquiring it, America has another interest in Iraqi oil. It would unquestionably like to split Iraqi oil from other sources. Doing so would weaken the Organization of Oil Producing Countries (OPEC) and lessen American dependence on Saudi Arabia and Russia. If it continues to control Iraqi oil, which is the cheapest in the world to produce, it could to some degree control world prices. To accomplish these objectives, it would like to increase production from about 2 million barrels a day in November 2004 to 8 million barrels a day in a decade. To reach such a goal will require tremendous capital investment. That, in turn, will depend in part on the ending of the insurgency. But if even a portion of that goal is reached, it could ensure the energy on which the Western economy depends. So the stakes are high. Attempting to accomplish these objectives through monopoly or imperial control would be anachronistic, unnecessary, and self-defeating. Wise statesmen will opt for the alternative, allowing the free play of market forces with production under Iraqi sovereignty.

Oil sales, of course, produce vast amounts of money. Control over expenditure of this money is almost as sensitive as control over the means of production. Regulations promulgated by the occupation authority give it this control. The authority has used it promiscuously. Nearly $19 billion of the $20 billion spent so far on contracts to rebuild war damage and run the administration has come from Iraqi oil revenues and frozen Iraqi bank accounts. Additional funds are to be borrowed against the future sale of Iraqi oil. At the same time, the authority has spent only about 3 percent of the $18.4 billion in American money allocated by Congress in the fall of 2003. Funds from both sources have been passed out primarily to American corporations such as Bechtel, Fluor, and Halliburton, without customary bidding procedures. At least part of these actions is of questionable legality. The May 2003 United Nations resolution, under which the occupation authorities have acted, required them to create an international oversight board. The occupation authority refused to do so for nearly a year and then dissolved itself before the audit report could be issued. These circumstances gave scope for various irregularities. The less significant but easiest to spot were overcharging. One firm, a subsidiary of Halliburton, the company of which Vice President Cheney had been chief executive officer, was found to have charged the Defense Department for 36 percent more meals than it actually provided (the company subsequently admitted to 19 percent overcharging). *Time* magazine reported on November 1, 2004, that the company also overcharged the government $61 million for fuel. Excessive charges by Halliburton and other companies are now being investigated. What is important, over the long run, in these events is not so much the actual waste of money but the setting of a style for the Iraqis. It will be difficult to convince Iraqi businessmen to be honest when such examples of fraud are placed before them by their American mentors.

At the center of the policy promulgated by Mr. Bremer and designed by the Bush administration was a series of moves that effectively "denationalized" the Iraqi economy. This policy was directed toward not only privatizing state-owned enterprises but also allowing them to be purchased 100 percent by foreign interests. The intent of the Bush administration policy was to make Iraq the perfect example of what the *Economist* called "a capitalist dream." Actually, it was not only that but more pointedly a foreign capitalists' dream. What the Bush administration and Mr. Bremer should have known was that the edicts establishing the new order were illegal under Security Council Resolution 1483. That resolution recognized the occupation authority but required it to abide by existing internationally binding conventions that were designed precisely to prevent occupying powers from "looting" the economies of the countries they had defeated in war. Even members of the group appointed by Mr. Bremer, all members of the governing council and the newly selected interim ministers, refused to implement the American edicts, and American corporations, undoubtedly under legal advice, refused to participate. But the image of American fairness and concern with law was severely damaged.

These moves were further compounded by policies promulgated to lift all restrictions on the importation of goods. These edicts were effected when the Iraqi economy was shattered by the war and so placed local entrepreneurs and manufacturers at a severe disadvantage. They simply could not compete, often lacking adequate machinery and access to raw materials, with cheap imported goods. Their inability, in turn, played a major role in causing nearly seven in ten male workers to be unemployed. As Naomi Klein has written,[*] "Bremer's reforms were the single largest factor leading to the rise of armed resistance" in Iraq.

In addition, American firms have been given the inside track on all the major reconstruction contracts, while most Iraqi firms and firms from other countries have been excluded. In December 2003, Deputy Secretary of Defense Paul Wolfowitz issued an order banning French, German, Russian, and even Canadian firms whose countries had opposed the American invasion. Such moves are as bitterly resented in Iraq, as they were in America when the British similarly treated the American colonists on the eve of the American Revolution.

Also harking back to the American Revolution was military policy in Iraq. If the occupation authorities knew their American history, they would have known that it was the presence of British troops in Boston that triggered the American Revolution.

---

[*] Her excellent article in the September 2004 issue of *Harper's Magazine* is entitled "Baghdad Year Zero: Pillaging Iraq in Pursuit of a Neocon Utopia."

Soldiers and civilians make poor neighbors. But, until the summer of 2004, they stationed American troops in Iraqi cities. Inevitably, the accumulation of small incidents, misunderstandings arising from the inability to speak one another's languages, and fear created pervasive hostility. Finally the high command realized the danger and largely pulled its troops out of cities into rural bases. Even had they not resumed the offensive against the cities in September and October, this would have been only a partial solution to the problem. Not only have the bases often been placed on top of or contiguous to Iraqi archaeological sites, and so have endangered or even already destroyed priceless cultural treasures, but in more recent memory, Iraqis see them as threats to their sovereignty. America has given them ample reason to believe this because it has made no secret of its plans to use Iraq as its main military base in the Middle East. Among the advantages put forward for this policy is that it removes an irritant in Saudi Arabian–American relations. But Iraqis, who remember how Britain similarly withdrew to remote bases in Iraq that it maintained long after "independence" and used them to overthrow an Iraqi government, will view American bases, as they viewed British bases, as daggers hung over their heads. Worse, having heard what the Neo-Conservatives in Washington have proclaimed, nationalists will see Iraq becoming an outpost for both American and Israeli policy.

Israeli policy is and will continue to be, under any foreseeable government, significant to Iraqi politics. Iraqis see Israel as a Western colonialist outpost in the Middle East. They are particularly frightened by the current Likud government of Prime Minister Ariel Sharon, but from the 1930s, they believed that Israel had been created by Britain to threaten the Arab world into doing what Britain wanted. Like other Arabs, Iraqis fear the massive Israeli program in weapons of mass destruction. And, finally, they sympathize with the Palestinians who have lost their homeland and are today living under what Israelis admit is a brutal occupation regime. These events and what the Iraqis see as the hidden hand of Western imperialism behind them constitutes a reservoir of anger, sense of impotence, and shame. It is partly because they hold the United States responsible—as the muscle and purse behind what they see as an aggressive Israeli policy— that America is so unpopular in Iraq. It follows that if the United States were to convene a fair, even-handed regional conference to air grievances and seek solutions to this festering wound, the effects in Iraq would be dramatic.

America today has two options in Iraq, stay or leave. "Staying the course," to use George Bush's term, is the one his administration has staked out as I write. America is continuing to try to establish security by fighting the insurgents; is continuing to control the oil industry and other major sectors of the Iraqi economy; and is "guiding" most other aspects of government through appointed "inspector-generals." American corporations continue to dominate the economy. Unemployment remains high because Iraqi businesses cannot compete with cheap imported goods. Then, if the government acts unacceptably, America (like Britain in the 1930s and 1940s) will probably either replace it or destabilize it so that it can be manned by "friendly" Iraqis. In this flow of events, trained and equipped by the Americans, the Iraqi army will again assume a power unbalanced by other institutions. Despite or even because of its training, the officer corps will again, as in the 1930s and 1950s, aspire to lead the nation. If it follows the likely pattern, it will rebel against the Americans and their local surrogates; if it does not, Iraqi nationalists, with or without army assistance, will continue to use terrorism because that will be their only available weapon to try to force America to leave.

At best, "staying the course" can be only a temporary measure as eventually America will have to leave. But during the period it stays, say the next five years, my guess is that another thirty or forty thousand Iraqis will die or be killed while the U.S. armed forces will lose perhaps five thousand dead and twenty thousand seriously wounded. The monetary cost will be hundreds of billions. Consider what the figures mean. Americans were horrified when about thirty-three hundred people were killed in the attack by al-Qaida terrorists on the World Trade Center on September 11, 2001. Iraq has already (at the time of this writing) lost about one hundred thousand during the American invasion and occupation.[*] In absolute terms that means that virtually every Iraqi has a parent, child, spouse, cousin, friend, colleague, or neighbor—or perhaps all of these— among the dead. More than half of the dead were women and children. In relative terms this figure equates to a loss in the very much larger American society of more than a million people.

It is not only the actual casualties that count. What wars of "national liberation" have taught us is that they brutalize the participants who survive. Inevitably such wars are vicious. Both sides commit atrocities. In their campaigns to drive away those they regard as their oppressors, terrorists/freedom fighters seek to make their opponents conclude that staying is unacceptably expensive and, since they do not have the means to fight conven-

---

[*] As reported by Dr. Les Roberts and a research team from Johns Hopkins University's Center for International Emergency, Disaster, and Refugee Studies, published in the medical journal *The Lancet* in November 2004 and summarized by the *New York Times* on October 29, 2004.

tional wars, they often pick targets that will produce dramatic and painful results. Irish, Jewish, Vietnamese, Tamil, Chechen, Basque, and other terrorists/freedom fighters blew up hotels, cinemas, nightclubs, and/or apartment houses. The more spectacular, the better for their campaigns. So the Irgun blew up the King David Hotel in Jerusalem in 1946; the IRA a Brighton (England) hotel in 1984; an Iraqi group the UN headquarters in Baghdad in 2003. Chechens blew up an apartment house in Moscow in 2003 while a Palestinian group blew up an Israeli-frequented hotel in Taba (Egypt) in 2004.

Faced with this challenge, the occupying power often reacts with massive attacks aimed at terrorists but inevitably also killing many civilians. To get information from those it manages to capture, it also frequently engages in torture. Torture did not begin at the Abu Ghuraib prison; it is endemic in guerrilla warfare. Two phrases from the Franco-Algerian war of the 1950s–1960s tell it all and ring true today: "Torture is to guerrilla war what the machine gun was to the First World War" and "Torture is the cancer of democracy." Guerrilla warfare and counter-insurgency inexorably corrupt the very causes for which soldiers and insurgents fight. Almost worse, even in exhausted "defeat" for the one and heady "victory" for the other, they leave behind a chaos that spawns warlords, gangsters, and thugs as is today so evident in Chechnya and Afghanistan. After half a century, Algeria has still not recovered from the trauma of its war of liberation against France. The longer the war in Iraq continues the more it will resemble the statement the Roman historian Tacitus attributed to the contemporary guerrilla leader of the Britons. The Romans, he said, "create a desolation and call it peace."

In carrying out this policy we can be sure of two things: the first is that Iraq will suffer grievously and the society that survives will be wounded, distorted, and far less than now likely to achieve a reasonably free and peaceful future. The second is that domestically American society will be angry, dispirited, and less democratic than today while internationally it will have lost much of the moral force that throughout its history, from the very Declaration of Independence, has been its most valued and most potent asset. In every sense and for both Iraqis and Americans, staying is, as was said of nuclear warfare, simply unacceptable.

The alternative, leaving Iraq, is not a single policy but presents two quite different varieties of conduct. One is what might be called "Vietnamization." In Vietnam, America sought to turn the war over the government and army of the South. Neither lasted very long. Since neither existed in post-Saddam Iraq, both a government and an army had to be created to give America the option to effect this policy. As I have written, few observers believe that either could long survive an American withdrawal. The best America might gain, if the process could be drawn out for several years, is a fig leaf to hide defeat; the worst, in a rapid collapse, would be humiliating evacuation, as in Vietnam.

The better form of "getting out," the second variety, involves choosing rather than being forced. Time is a wasting asset; the longer the choice is put off, the harder it will be to make. The steps required to implement this policy need not be dramatic, but the process needs to be affirmed. Thus, initial steps could be merely verbal. America would have first to declare unequivocally that it will give up its lock on the Iraqi economy, will cease to spend Iraqi revenues as it chooses, and will allow Iraqi oil production to be governed by market forces rather than by an American monopoly. If an American administration could be as courageous as General Charles de Gaulle was in Algeria, when he admitted that the Algerian insurgency had "won" and called for a "peace of the brave," fighting would quickly die down as it did there and in all other guerrilla wars. Then, and only then, could elections be meaningful. In this period, Iraq would need a police force but not an army. A UN multinational peacekeeping force would be easier, cheaper, and safer than creating an Iraqi army which in the past destroyed moves toward civil society and probably would do so again, probably indeed paving the way for another Saddam Husain.

A variety of "service" functions would then have to be organized. Given a chance, Iraq could do them mostly by itself. It would soon again become a rich country and has a talented, well-educated population. Step by step, health care, clean water, sewage, roads, bridges, pipelines, electric grids, housing, etc., could be mainly provided by the Iraqis themselves, as they were in the past. When I visited Baghdad in February 2003 on the eve of the invasion, the Iraqis with whom I talked were proud that they had rebuilt the Tigris bridge and other facilities that had been destroyed in the 1991 war. They can surely do so again.

In its own best interest, the Iraq government would empower the National Iraq Oil Company (NIOC) to award concessions by bid to a variety of international companies, each of which and NIOC would sell oil on the world market. Contracts for reconstruction paid for by Iraqi money would be awarded under bidding, as they traditionally were, but to prevent excessive corruption perhaps initially supervised by the World Bank. Where other countries supplied aid, they could be given preferential treatment in the award of contracts as is common practice elsewhere. Where World Bank loans are involved, the Bank would follow its regular procedures. Abrogating current American policies that work against the recovery of Iraqi industry and commerce would spur development since any reasonably intelligent and self-interested government would emphasize getting Iraqi enterprises back into operation and employing Iraqi workers. That process could be speeded up through international loans, commercial agreements and protective measures so that unemployment, now at socially catastrophic levels, would be diminished. Neighborhood participation in running social affairs and providing security are old traditions in Iraqi society and allowing or favoring their reinvigoration would promote the excellent side effect of grassroots political representation. As fighting dies down, reasonable security is achieved and popular institutions revive, the one million Iraqis now living abroad will be encouraged to return home. In the aggregate they are intelligent, highly trained, and well motivated and can make major contributions in all phases of Iraqi life.

In such a program, inevitably, there will be setbacks and shortfalls, but they can be partly filled by international organizations. The steps will not be easy; Iraqis will disagree over timing, personnel, and rewards, while giving the process a chance will require American political courage. But, and this is the crucial matter, any other course of action would be far worse for both America and Iraq. The safety and health of American society as well as Iraqi society requires that this policy be implemented intelligently, determinedly and soon.

INDEX

ABOUT THE AUTHOR

WILLIAM R. POLK studied at Harvard (where he earned his B.A. and Ph.D.) and read Arabic and Turkish at Oxford (where he earned his B.A. and M.A.). As a fellow of the Rockefeller Foundation, he also studied in Iraq and other parts of the Middle East. He taught Middle Eastern history and politics and Arabic at Harvard until 1961, when he became a member of the Policy Planning Council of the Department of State, responsible for the Middle East and North Africa. In 1965 he resigned to become Professor of History and Founding Director of the Center for Middle Eastern Studies of the University of Chicago. He was also a founding director of the American Middle Eastern Studies Association. He has lectured in more than a hundred universities and colleges as well as at the Council on Foreign Relations, the Canadian Institute of International Relations, the Royal Institute of International Affairs, and the Institute of World Economy and International Affairs of the Soviet Academy of Sciences. Among his many books are *The United States and the Arab World, The Arab World Today, The Elusive Peace: The Middle East in the Twentieth Century,* and *Neighbors and Strangers: The Fundamentals of Foreign Affairs.* His articles and essays on subjects tangential to this book can be accessed on his Web site, www.williampolk.com.